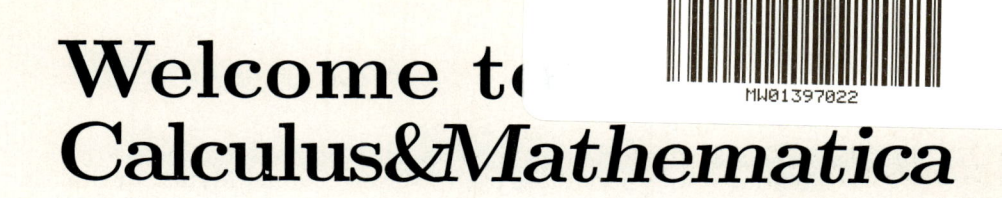

Welcome to Calculus&Mathematica

Bill Davis
Ohio State University

Horacio Porta
University of Illinois, Urbana-Champaign

Jerry Uhl
University of Illinois, Urbana-Champaign

with special assistance from
Alan Deguzman, Justin Gallivan,
Corey Mutter, and David Wiltz

Addison-Wesley Publishing Company

Reading, Massachusetts • Menlo Park, California • New York
Don Mills, Ontario • Wokingham, England • Amsterdam • Bonn
Sydney • Singapore • Tokyo • Madrid • San Juan • Milan • Paris

Reprinted with permission from "Moving Beyond Myths: Revitalizing Undergraduate Mathematics," © 1991 by the National Academy of Sciences. Courtesy of National Academy Press, Washington, DC.

Peter Lax, Robert White, Ronald Douglas, and Lynn Arthur Steen are quoted by permission of the Mathematical Association of America.

This material was prepared with partial support from National Science Foundation Grant Nos. USE 9053372, DUE 9252484, and USE 9153246 at the University of Illinois and The Ohio State University. However, any opinions, findings, conclusions, or recommendations expressed herein are those of the author(s) and do not necessarily reflect the views of NSF.

None of Addison-Wesley, Wolfram Research, Inc., or the authors of Calculus&*Mathematica* makes any warranty or representation, either express or implied, with respect to the Calculus&*Mathematica* package, or its performance, merchantability or fitness for particular purposes.

None of Addison-Wesley, Wolfram Research, Inc., or the authors of Calculus&*Mathematica* shall be liable for any direct, indirect, special, incidental, or consequential damages resulting from any defect in the Calculus&*Mathematica* package, even if they have been advised of the possibility of such damages. Some states do not allow the exclusion or limitation of implied warranties or consequential damages, so the above exclusion may not apply to you.

Mathematica is a registered trademark, and *MathLink* and *MathReader* are trademarks of Wolfram Research, Inc. *Mathematica* is not associated with Mathematica, Inc., Mathematica Policy Research, Inc., or MathTech, Inc.

Many of the designations used by manufacturers and sellers to distinguish their products are claimed as trademarks. Where those designations appear in this book, and Addison-Wesley was aware of a trademark claim, the designations have been printed in initial cap or all caps.

Copyright © 1994 by Addison-Wesley Publishing Company, Inc.

All rights reserved. No part of this publication may be reproduced, stored in a retrieval system, or transmitted, in any form or by any means, electronic, mechanical, photocopying, recording or otherwise, without the prior written permission of the publisher. Printed in the United States of America.

ISBN 0-201-58463-8

1 2 3 4 5 6 7 8 9-CRS-9897969594

To Jo Dee and Piki.
These terrific people manage to get along with all three of us.
Imagine that!

Did You Ever Think?

Did you ever think that the math courses that most of you have taken in the past did nothing more than to try to program you to do rote calculations automatically?

Did you ever think that you forgot the rote calculations you learned soon after the exam?

Did you ever think that these courses treated you as little more than a human computer?

Did you ever think that nobody ever told you what all those rote calculations were good for and what they have to do with life around you?

Did you ever think that you were often told to learn something because "You'll need it in the future," but then you never saw it in the future?

Did you ever think that, as a former president of the American Mathematical Society said, "Calculus, as currently taught, is full of inert topics"?

Did you ever think that one of the inventors of calculus said, "It is unworthy of excellent persons to lose hours like slaves in the labor of calculation"?

Did you ever think that the software program *Mathematica*, running on a little computer on your desk, can do for you all the rote calculations you were ever taught?

Did you ever think that the software program *Mathematica* can do for you all the graphs you, your teachers, or your books ever did?

Did you ever think that it is possible to learn mathematics with the help of a book that's electronically alive and in which every example and plot can be changed and reexecuted as you see fit?

Did you ever think that it is possible to learn more mathematics by pictures than by memorization of words that seem to have no meaning?

If any of this piques your curiosity, then you're likely to learn a lot in Calculus&*Mathematica*.

Contents

About This Booklet	vi
What Is Calculus&*Mathematica*? How Has It Been Received by Its Students?	1
Lesson Interdependence	8
Getting Going	9
Calculus&*Mathematica* with Macintosh	14
Calculus&*Mathematica* with NeXTStep	31
Calculus&*Mathematica* with Microsoft Windows	41
How to Type Using the Calculus Font on Macintosh, NeXTStep, and Windows Platforms	53
Moving Files Among Different Kinds of Machines	64
Trashed Notebooks	66
Glossary of Computerese	71
Why Was Calculus&*Mathematica* Written? What Principles Underlie the Course?	74
Acknowledgments	80

About This Booklet

The sections on getting started on the different platforms were written by the original Development Team, Alan Deguzman, Justin Gallivan, David Wiltz, and Corey Mutter. We must say more about the contributions these people have made to this project. It wouldn't be in the condition it is if it weren't for them. The original team at Illinois made the lab work. Following in the footsteps of Don Brown, they have installed and maintained machines and network. They still maintain the labs and the networks in the labs there, even as the hardware changes. Alan Deguzman has been a lab assistant and is the Macintosh systems administrator in the Illinois lab. He is a presenter and co-presenter of Calculus&*Mathematica* at many conferences and shows, and he wrote `CMForm[ ]` while helping Bill Davis at Interactive Math Textbook Project workshops in Seattle and Houston. He is an extraordinary teacher in such workshops. Justin Gallivan was the pioneer who believed that these lessons really would run on Windows machines as soon as WRI had a beta version of *Mathematica* in the works for Windows. He had to make a version of the Calculus&*Mathematica* font work on Windows machines, and he realized that typing would have to be different on Windows machines, so he wrote the first successful typing program for the Calculus&*Mathematica* `KeyInKey`. He wrote the installer for Windows and is in charge of making it work. He has also presented Calculus&*Mathematica* at various conferences, and is the person who is most responsible for our running hands-on workshops on Windows machines. Dave Wiltz is our systems administrator, and our expert on the system-level work that must be done on the mix of platforms. He took Justin's first font for Windows and made it work better. Dave inherited the Calculus&*Mathematica* Notebook to TeX conversion program maintenance from Don (Chip) Brown. In fact, he used that program, called the `Chipper`, to convert from earlier fonts to the current Calculus font. He is an expert on all sorts of technical stuff, like compression programs, file manipulation, and the like. Of course he is an evangelist for Calculus&*Math-*

ematica, and a big help with presentations and workhops. Corey Mutter joined the team a couple of years ago with a flourish; he wrote **CMType** for Windows machines, which replaced **KeyInKey**. Now he takes care of moving the lessons from our development machines to Windows and making them look good there. He is the OSU lab's Calculus&*Mathematica* guru.

From the Development Team

The Calculus&*Mathematica* Development Team would like to take this opportunity to thank Bill Davis, Horacio Porta, and Jerry Uhl for following through on their vision of a calculus revolution. We'd also like to thank them for allowing us to join in for the ride (and what a ride it has been) and, most especially, for teaching us some of the subtle pleasures of life. We are grateful for the chance to work with them on this project.

What Is Calculus&*Mathematica*? How Has It Been Received by Its Students?

Here is a description of Calculus&*Mathematica* taken from the National Research Council report *Moving Beyond Myths*:

An innovative calculus course ... [which] uses the full symbolic, numeric, graphic, and text capabilities of a powerful computer algebra system. Significantly, there is no textbook for this course—only a sequence of electronic notebooks.

Each notebook begins with basic problems introducing the new ideas, followed by tutorial problems on techniques and applications. Both problem sets have "electronically active" solutions to support student learning. The notebook closes with a section called "Give It a Try," where no solutions are given. Students use both the built-in word processor and the graphic and calculating software to build their own notebooks to solve these problems, which are submitted electronically for comments and grading.

Notebooks have the versatility to allow reworking of examples with different numbers and functions, to provide for the insertion of commentary to explain concepts, to incorporate graphs and plots as desired by students, and to launch routines that extend the complexity of the problem. The instructional focus is on the computer laboratory and the electronic notebook, with less than one hour per week spent in the classroom. Students spend more time than in a traditional course and arrive at a better understanding, since they have the freedom to investigate, rethink, redo, and adapt. Moreover, creating course notebooks strengthens students' sense of accomplishment.

Calculus&*Mathematica* presents a complete rethinking of:

→ the mathematics of calculus

→ calculus as a first course in scientific measurement

What Is Calculus&*Mathematica*? How Has It Been Received by Its Students?

→ mathematics as an empirical science
→ how to present calculus ideas visually
→ what students do in calculus
→ what motivates students to do calculus
→ helping students realize that calculus is a discipline connected with reality
→ how to give students a sense of accomplishment
→ how to motivate students to write about calculus
→ how a math class should be conducted
→ how technology should be used in mathematics education.

Unlike any other calculus course, Calculus&*Mathematica* attempts to involve you actively in your own learning by putting you in a position to acquire mathematical ideas visually. This means that instead of hanging technology on the end of the your learning process, Calculus&*Mathematica* uses technology to initiate your learning process. It accomplishes this through its electronic interactive text, as described above, and it invites you, the student, to write about the mathematics you are doing as you are doing it. Typically, Calculus&*Mathematica* students learn by doing at the machines, attend no lectures, but spend a lot of time talking to each other.

Here are some unedited student reactions to Calculus&*Mathematica*. You can expect your Calculus&*Mathematica* experience to be like theirs.

I've been studying math for years now and doing pretty well at it. But I never knew what I was studying before you got me on this computer and I could see it.

I have never seen anything taught in a manner that so frees the student from the unnecessary baggage that goes along with the important stuff.

This was the first calculus class that I have ever taken and enjoyed.

If you put in any amout of time with a little bit of effort, you can't help but learn.

Calculus&Mathematica makes all other homework seem boring.

It gave me the real feeling that I was actually doing something, not just plugging and chugging away.

In other classes you had to be good at following recipes, but [in Calculus&Mathematica], . . .I got to think about the recipes . . .what I am trying to say is that Calculus&Mathematica piqued my interest in math. I really mean that.

I think that this was my best class this year. I am walking away from it feeling like I actually learned something useful that I might actually use some day.

[Calculus&Mathematica] takes the boredom and frustration out . . . and let's you be creative and learn on your own—at your own pace.

This course . . .has an advantage over textbook courses in that I am given a lot more freedom in how I want to learn.

Calculus&Mathematica is the only way I've seen to teach any math higher than geometry in such a fashion that students gain a fundamental understanding of the concepts involved. In a normal class, the opportunity to discuss how and why mathematics works gets lost in all the drudgery and rote calculations. Mathematica takes care of the drudge calculations and leaves the student free to concern himself primarily with the essential concepts that truly make up math.

When this class first began, I felt as if I wasn't learning as much as the folks reading that nasty blue book. I felt this way a long time and was worried that if I took a regular math class in the future I would be FUBAR (an acronym). My revelation came when one of the blue people had a question about something . . . . I mumbled I don't know if I can answer your question . . . because we do different things in my class. But, I was really thinking . . . . Hey . . . those lousy computers do all the work so I don't know squat. But he asked me the question, and to my surprise, I knew the answer. Plus, I knew the why of the answer because that's what this class does. I definitely prefer [Calculus&Mathematica] to the blue method.

I think that C&M has helped me mature intellectually. I think about problems in a better fashion now, perhaps more ordered and more logically cohesive.

The C&M program definitely taught me a lot about how I learn and how I get things done, and that I actually do have the ability to learn what is happening in a math class.

I have become interested in calculus and no longer fear it to be the monster that I once thought that it was.

Of all the lessons we did . . . , I can't really think of one where I don't (at least now) know what was going on and what was happening, as opposed to (traditional Calculus I) where I was totally lost the entire time and lucky to escape with my life.

C&M has brought to me something I thought I would never have—a bit of understanding of the important relationships in mathematics. Some people (say in my [traditional] calc I class) were able to look at the texts and seem to understand, but I have been frustrated very many times by not being able to. I guess that's why I signed up for this class. I used to not like math. But I was just being a product of the way I was taught.

I have always been extremely bad in anything mathematically intensive. This class actually allowed me to realize that mathematics (at least C&M) doesn't have to be a huge chore, but can actually be beneficial and a major learning experience.

I like the "hands on" type learning, rather than some boring professor lecturing at you. That way seems rather silly to me. Here if you don't get it right the first time, you sit here (at the machine) until you do get it right, therefore learning in the process . . .

Lectures are boring because I have a short attention span. I tend to sleep or write poetry during lectures and get nothing out of them. [Calculus&]Mathematica captures me a lot more. Plus, it is easier to learn on my own time.

Actually, to come straight out, I loved this class. Hey, I don't mean to brown nose or anything, lemme tell ya why. Last semester I took Calc II without the Mathematica option. I hated it. I flunked. I mean, I was ticked. I. . .didn't need this slop . . . . I was just NOT interested in the class. I could care less what I learned. So, this is where Calculus&Mathematica comes in. For starters, this is way more interesting!! For example, it is 1:20AM, and I have a Chem final tomorrow morning that I need to study, and I was here last night rather late also. Why? Well, I don't really "love" it THAT much, but, hey, I feel like getting the work done. I like playing around with the functions, learning new ways to do things. I mean, I even remember most of the stuff that we "learned". The ways that problems are presented are intuitive and fairly easy to grasp, once the puzzle is put together . . . . That helps a lot in the learning process.

In this course, I was allowed the freedom to prove something to myself, if I wasn't convinced. . . . Perhaps the most important thing that happened was that I started to think about what formulas said. You have to do that in this course, unlike the other classes I've had.

I felt fully enriched by this course. This course . . .was filled with surprises around every turn. I discovered new applications of the [traditional] Calculus I had learned, and had my mind challenged by the many interesting problems in this course.

I dreaded taking Calc II. I saw it as a bunch of formulas I needed to know to get through Calc III. [Calculus&]Mathematica certainly made it interesting . . .I don't think I am lacking anything as far as comparing it to a standard calc class. I am able to do things on paper and able to read a calc book and understand, more so than a few engineers who got A's in (traditional) Calc II. . .

The set up [of the electronic text] on the computers gave a lot more freedom and independence. I felt that I decided when and how I learned the material which made it easier on me. If I understood the basics and tutorials on the first time through, I could go right to the homework. If I was having trouble, I could go back and work through a few more examples. I thought the homework . . .was very effective in making sure you understood the important general concepts of the course. I did not miss all the mindless number crunching stuff my friends in the conventional course had to go through.

Before I came to college I loved math. It had always been my favorite class all through school. When I graduated high school I even got the math award. When I got to [college] my love of math quickly died . . .I no longer understood math or calculus for that matter. Then I switched majors and I had to continue taking more Math. Last semester I was also in [Traditional Calc II] and was getting a D- before I dropped it. When I went to [registration] this semester just to change times and someone suggested I take [Calculus&]Mathematica. I figured it couldn't be any worse than last semester so I signed up for it. I'm glad I did. I wouldn't say that I love math now, but I am beginning to like it again. This class not only helped teach mathematics but it explained mathematics. The normal classes don't do this.

Now, I look at math in a different light. I was helping a friend in the normal (read: old-fashioned, obsolete) section, and I worked a problem down to the integral and stopped there, satisfying myself that a computer could take it from there. My friend looked at me, stunned that I had not done the hardest part of the problem and considered myself finished. I was finished, for I had done the thinking behind the problem and didn't want to bother myself with the petty details of working through memorized procedures. That's what a computer's for.

I took Calculus&Mathematica because I like the fact that we have electronic notebooks which means we have infinite sets of examples. I cannot learn from one example such as in the [traditional course]. I would rather be in this lab and learn things on my own, by trial and error, than to run to a tutor who will only program my memory to do only certain kinds of problems.

The thing I value most about Mathematica is the graphing. Simple things like the unit circle, sine and cosine, exponential growth—only this semester have I felt I have an idea of what they all mean, and I have so much more to learn. I can actually explain things to others. I have never felt good about math before.

It always gave me great satisfaction to look at a finished product on the screen, knowing that doing it by hand would have been incredibly painful, if not impossible.

I didn't get stuck up in all of the pencil work that it usually took in traditional calculus. I was able to grasp the concepts a lot better because of it. Overall I really enjoyed it.

When we first began working with this crazy machine I found myself wondering what I had gotten myself into. Soon I came to love this wonderful and powerful tool called Mathematica.

The first few weeks of this course were stress filled for me. I had never really had any [computer] experience, so the whole system was totally foreign to me. Then there was the Mathematica program itself. I had no idea how this thing worked. Doing my first few homework assignments was very frustrating. There was a time when I seriously considered changing to a normal section . . . . But, I stuck with it, and I am very happy I did.

I have enjoyed Mathematica because it allows me to satisfy my natural curiosity of what is happening in functions without forcing me to go insane due to having to deal with monstrous equations.

I consider mathematics to be my weakest discipline. One thing that does help me a lot is visualization of functions. Calculus&Mathematica has helped me tremendously because graphs of functions are readily available and easily mutable. I think that manipulation of graphs should be an essential part of all calculus classes. They personally help me more than any other medium.

I found that we have better learned what we are doing in this class than the standard Calc II class. More than likely this can be attributed to the fact that we do not have to learn the ugliness of all the formula manipulations that the other class does.

Instead we are left the task of understanding and satisfying our curiosity as to what is really going on with the functions. It is much easier to change a formula around and see what the changes do when you can graph the old and new function and not have to do hundreds of calculations to get these graphs. As proof of this our knowledge of expansions, while at first glance is seemingly trivial, came in very useful in more than one instance. Last night Marc and I were explaining how to do a relatively easy substitution into an expansion to a regular student (in traditional Calc II). While we found it very easy because we had the opportunity to play with them for so long, the other student was very confused and we had to explain it several times before we got him to understand it.

One thing that really surprised me is the focus on communication. In [precalculus], the important thing was making the answers right by following the equations. [C&M], on the other hand, really encouraged me to explain my results using real-world English, rather than just a sequence of mathematical expressions and terms . . . . And often times this stressing of communication has even helped when my answers were wrong - either I explained the process by which I arrived at my wrong answer, and therefore made tracing the error easier; or I explained why I was unable to get what I thought to be the "right" answer.

I feel 100 times more confident about my own math skills, or at least in my problem solving ability. (I can always look in a book if I forget the specifics, but I now feel like I can use this stuff in ways that I never would have thought of before, things that are not carbon copy book problems.). . . I have started to notice aspects of one class carrying over to another. Similarities in fields I thought unrelated before. An interconnection between math and language and programming and everything just kind of fits together a little better now.

[C&M] has been a God-send for me. I have been diagnosed with dislexia and long term memory retrival problems. I took [Calc II] by the regular lecture way and it was a disaster. I went to every class, resitations, and got tutors. I failed horribly. The instructor was arogent and felt that the class was not a mathematical challenge for her! . . . I then took [C&M] and have gotten A's and B's. I now understand calculus! I hardly have to study . . . . The computer tells me in detail how it does something and it does not care how many times it has to tell me. This is the way to learn . . . . I want to learn and need to learn. [C&M] is the way to do it. I would strongly advise the math department, along with other departments to look at this teaching method and use it to teach as many classes as possible, giving students a true alternative to a failing traditional method of education!

As much as Luke Skywalker loved his time in math he had always known something was wrong . . . . His math abilities grew and grew, but he always felt that the methods of teaching were a constraint on his ability to really understand the mathematics being taught. When he had grown old enough he finally met Obiwan Kenobee who told him that he was meant to be a math master and that the slight force he had always felt would eventually become one with him and he would understand all. Obiwan explained to him that all of this would begin with the destruction of the old math empire .... In the next year Luke created a calculus program on computer and began slowly defeating the dark side of math that had always haunted him. As time went on he transferred all the mathematics the rest of the world could comprehend on to a computer and defeated the empire of long boring lectures and unnecessary computation. The force prevailed and true mathematics finally came through.

Most of these comments came from students who had taken at least one traditional calculus class before entering Calculus&*Mathematica* sections conducted by teaching assistants. The remaining comments are from university, community college, or high school students enrolled in Calculus&*Mathematica* sections.

Lesson Interdependence

If you want to design your own calculus course using Calculus&*Mathematica*, you should pay some attention to the way the different lessons depend on each other. Here's a quick summary.

If you do all of 1.xx series of Lessons (Growth) then you are ready to undertake the 2.xx series of lessons (Integration).

If you do all of 1.xx series of Lessons (Differentiation), and you do Lessons 2.01, 2.02 and 2.03 (Integration), then you are ready to undertake the 4.xx series of lessons (Approximation). Most students who have taken a year of high school calculus or one semester of college calculus should be ready to indertake the 4.xx series of lessons (Approximation).

If you do all of 1.xx series of Lessons (Differentiation), and you do Lessons 2.01, 2.02, 2.03, 2.04 and 2.05 (Integration), then you are ready to undertake the 3.xx series of lessons (Vector Calculus). Most students who have taken a year of high school calculus or one semester of college calculus should be ready to indertake the 3.xx series of lessons (Vector Calculus). Students moving from the traditional course into Calculus&*Mathematica* Vector Calculus should do lessons 1.09 and 2.05 at the very beginning of Vector Calculus.

Getting Going

■ **Before you start**

You'll need a computer capable of running *Mathematica* with the Notebook feature. All but the lowest level Macintoshes should be fine. Most 486 machines running under Windows or NeXTStep will do quite well. Any NeXT computer will do. In the near future, any X-Windows work station will be great. The memory of some of these computers might have to be upgraded to run *Mathematica*.

You'll need a copy of *Mathematica*. You can order *Mathematica* by calling 1-800-441-MATH or 217-398-0700. Tell them what machine you're going to use, ask them how much memory you need, and tell them you are going to use it to take the Calculus&*Mathematica* course. As a *Mathematica* student, you should qualify for the lower-priced student edition of *Mathematica*.

You'll not need to know how to use *Mathematica*. Instead of taking a separate course in *Mathematica*, you'll learn *Mathematica* gradually as you need it in the course.

You'll need the software that comprises the electronic text. You must have *Mathematica* on your machine before you can use the Calculus&*Mathematica* software. If you bought the whole boxed package, you have all you need. To order the software or the printed supplements, call Addison-Wesley at 1-800-447-2226. The special font, which comes with the software, must be loaded into your machine before you can begin.

You'll not need to have a strong background in algebra and trigonometry. If you're not so good at pushing around the x's and y's you see in algebra, don't worry. *Mathematica* will do for you almost all the algebra you need. Of course, you have to know enough about algebra to manage *Mathematica*. If you've never heard of solving equations or factoring, then you probably don't know enough algebra. If you haven't had much trigonometry, don't worry because you don't need much trigonometry to succeed in Calculus&*Mathematica*.

You'll need to be open-minded enough to realize mathematics is lots more than memorized calculation procedures. This is not the traditional calculus course. If you throw yourself into it, this course will build long-lasting knowledge, not just memorized procedures that you learn and then forget. You won't be asked to memorize a lot of procedures because the computer will handle them for you. This is a highly visual course that allows you to see what's going on by putting you in the position of making math happen. You'll learn by doing and by discovery. You'll learn what calculus is and what it's good for in a way not seen in other math courses.

You'll need someone else to work with at least some of the time. It might be difficult to learn in isolation. Contrary to what you might believe about mathematics, learning mathematics is a highly social activity that does not lend itself to learning in isolation. If you're a home schooler, find someone else to work and talk with. Two of you working together can learn a lot more than two of you working separately.

Above all, you'll need the desire to learn calculus. This is not a course for those who have a half-hearted commitment to learning. If you are willing to do the work, you'll learn.

▪ Starting out in Calculus&*Mathematica*

Load the Calculus&*Mathematica* software into your computer as discussed later in this booklet, making sure that the special Calculus Font is also loaded. If you have no prior *Mathematica* experience, open up the folder called "Feel of *Mathematica*" and play with it for a couple of days. Then open up the folder called "1.Growth" and start to work with Lesson 1.01. At this point, you will be ready to begin to understand the advice given below.

▪ Tips on using Calculus&*Mathematica* for your learning

> Refer to these tips from time to time as you progress through the course.

Each Calculus&*Mathematica* lesson consists of four parts:
    Basics: problems with answers,
    Tutorials: more problems with answers,

Give It a Try: problems waiting for machine answers,
Literacy Sheet: problems waiting for hand answers.

The Basics deal with the basic ideas of the lesson. The Tutorials deal with sample ways of using the basic ideas. The Give It a Try gives you the chance to use the basic ideas electronically. The Literacy Sheet signals the level of problem that you should be able to handle away from the machine in bull sessions around a cup of coffee or other beverage.

Using the electronic lessons. Take a Give It a Try problem that you want to solve. If you can solve it without using the Basics or Tutorials, go ahead and solve it. If you cannot solve it, then shop around in the electronic Basics and Tutorials files trying to find an idea or a technique that you can bring to bear on the problem. *The Give It a Try problems are the heart of Calculus&Mathematica.*

Using the Basics and Tutorials. After you have been working on the Give It a Try problems for a while, peruse the Basics and the Tutorials to see whether you've missed something you're interested in. It's always good to have a perspective about what's in the Basics and Tutorials. If you skip them completely, then you're likely to spend hours trying to do something that would have been simple had you spotted a certain fact or idea in the Basics or the Tutorials. *Think of the Basics and the Tutorials as research sources for doing the Give It a Try problems.* As you are going through a problem from the Basics or the Tutorials, play with the problem by changing functions or numbers and rerunning. In this way, almost every example becomes as many examples as you want.

Using your eyes. One of the beauties of learning the Calculus&Mathematica way is your opportunity to learn through graphics you can interact with. *In this course, your eyes will send ideas directly to your brain.* And this will happen without the distraction of translating what your eyes see into words. Take advantage of this opportunity to learn visually with pure thought uncorrupted by strange words. The words go onto an idea only after the idea has already settled in your mind. This aspect of Calculus&Mathematica distinguishes it from other math courses.

Lots of students who have had trouble learning math because they were bad at symbol pushing or had a hard time with vocabulary have done very well in Calculus&Mathematica.

Using the printed supplement to support the electronic lessons. It's not a good idea to try to acquire new ideas by trying to read the printed supplement first. Few students have ever learned mathematics by reading a printed book, and the chances are that you will not be able to learn only by reading the printed supplement. *Your first exposure to a new idea should be on the live computer screen where you can interact with the Basics and the Tutorials to your own satisfaction.* What you see on the live screen doesn't carry the same power when it appears on the static printed page because you can't interact with the printed page.

The printed supplement becomes handy when you want to browse the course or review what you've learned from the electronic lessons. It's also a handy place to find locations of sample *Mathematica* code. The printed supplement is also handy because the Literacy Sheets live there.

Memorization of rote calculation procedures. You know very well that you use memorization as a last resort. *Memorization has never been the path to learning.* In other math courses, students resort to memorization because many of the topics under study mean nothing to their intellects.

Calculus&*Mathematica* strokes your intellect by setting you in the midst of computer-generated visualizations that you are not likely to forget after you interact with them on the live screen. Also, in this course you learn by doing and by discovery. You get the chance to announce mathematical ideas in your own words. All this results in long-lasting knowledge, not just short-lived memorized procedures for calculation that you forget soon after the exam.

Copying, pasting, and editing. Copy, paste, and edit as much as you can. If you can find in the electronic Basics and Tutorials some *Mathematica* code or text that you want to adapt to your answer to a Give It a Try problem, then copy it, paste it, and edit it to suit the problem you are working on. And don't feel that you have to apologize for this. *There is no reason to try to tough it out when you can copy, paste, edit, and then go on to a new challenge.*

Many Calculus&*Mathematica* Give It a Try problems lend themselves to the copying, pasting, and editing approach. These problems are present to encourage you to become familiar with the ideas in the Basics and the Tutorials. You should knock off these problems first. Many other problems do not lend themselves to copying, pasting, and editing. After you are finished with the copying, pasting, and editing problems from a given lesson, you are likely to have enough experience to tackle the Give It a Try problems that don't lend themselves to this method.

Using *Mathematica*. Some new Calculus&*Mathematica* students feel that they must study *Mathematica* as a programming language before they can begin their study of Calculus&*Mathematica*. These students have condemned themselves to needless misery. In Calculus&*Mathematica*, you learn as much *Mathematica* as you need on a just-in-time-basis. You learn it gradually and always in context. The lessons themselves contain all the *Mathematica* code you need in context ready for you to copy, paste, and edit when and if you see fit. The going might be a bit rough for the first couple of weeks as you adapt to an unfamiliar environment, but if you're like 90% of Calculus&*Mathematica* students in the field tests, you'll become comfortable by the third week.

Remember: Calculus&Mathematica is a math course and not a programming course. After all, what programming course would encourage copying, pasting, and editing?

Using *Mathematica*'s word processor. When you work the Give It a Try problems, use *Mathematica*'s word processor to write the reason you are doing an upcoming plot or calculation. After the plot or calculation is done, use *Mathema-*

tica's word processor to write what was revealed. Take pride in your writing and use words you feel comfortable with. If you can write about something, you've got a handle on it. *Writing about calculus is an important part of your learning process*, and it's fun. One Calculus&*Mathematica* student put it this way: "It always gave me great satisfaction to look at a finished product on the screen."

Using the Literacy Sheets. Use the Literacy Sheets after you have completed the assigned Give It a Try problems and have some machine experience with the ideas of a lesson. It's O.K. to use the machine or to look in the book to find help in answering the questions in the Literacy Sheets. *After you are done with a lesson, you should try to answer the Literacy Sheet questions orally or with pencil and paper only.* This is a good way of putting some polish on what you already know.

Get together with a few others to have Literacy Sheet sessions to discuss the math of a lesson. Your knowledge is no good unless you can discuss it with someone.

Using pencil and paper. Many new Calculus&*Mathematica* students have the notion that they must be able to do something by hand before they have the right to do it by machine. These students are wrong. Calculus&*Mathematica* puts you in the position of learning visually through graphics and automatic calculations. You cannot learn visually by doing hand calculations. Once you are free of the yoke of hand calculation, you are free to understand the math behind calculus and to find out what calculus is good for. That's what this course is all about. *Hand calculations, as reflected in the Literacy Sheet problems, come into play after you've used the machine in your learning.* Simple hand calculations do have their place in bull sessions away from the machine.

Experience has shown that Calculus&*Mathematica* students become more proficient at hand calculations than you might predict. In fact, a study done by the College of Education at the University of Illinois found that it is impossible to distinguish Calculus&*Mathematica* students from traditionally trained students on the basis of their ability at hand calculation!

Using your friends. Calculus&*Mathematica* students who work a lot with others usually do very well. Calculus&*Mathematica* students who always work alone are usually either the very best students or they are in big trouble. *Unless you are a mathematical genius, try to work with others.* And if you are a mathematical genius, you'll be surprised at how many friends you can make by working with others. Either way, you'll be happy when you work with others. Ask your instructor whether group work will be accepted. It's a good deal for the instructor because group work results in less grading time.

Using your own ideas. You are the best judge of what is working for you. Try different things. Play with this stuff. Develop your own ideas about using Calculus&*Mathematica*. All of the best ideas about learning calculus this way have come from students. That's you.

Calculus&*Mathematica* with Macintosh

**by Alan DeGuzman,
Calculus&*Mathematica* Development Team**

Congratulations! You have selected Calculus&*Mathematica*, the most exciting and complete interactive calculus text available. This booklet will tell you everything you need to know to get you acquainted with the Macintosh and will describe features of the *Mathematica* program.

This is intended to be a learn-by-doing guide. To get the most from it, you should be sitting at a computer and actually doing the operations that are laid out for you. Don't be afraid to experiment on your own as well. One of the virtues of Macintosh is its simple, intuitive interface.

A note about the pictures: Don't be too concerned if the pictures shown here don't exactly match what shows up on your screen. Your system set-up will probably be a little different. If you've customized your set-up in any way, you probably won't need to read the introduction anyway.

If you are a beginner, don't be intimidated by the terminology. Everything is clearly spelled out, and there are plenty of illustrations. By reading through this, you will pick up a lot of the concepts quickly and easily.

Notes on reading this guide

Keep these few things in mind while reading this guide:

→ References to selections in the menus or the items in the menu bar themselves will be printed in this style: **File**, **Edit**, **Print**, etc.

→ References to disks, folders, applications, or other icons on your desktop will be printed in this style: **Macintosh HD**, **Trash**, **Mathematica**, etc.

→ Some pictures have Help balloons for further explanations. These are *not* the same as System 7 Help balloons. They are used only to help explain the illustrations.

→ All pictures in the manual show Calculus&*Mathematica* running under System 7 (more specifically, 7.1). System 6 users should understand that *Mathematica* 2.2 won't run under System 6 unless they order a special version from Wolfram Research. Apple Computer Inc. does not support System 6 anymore, and System 7 offers a host of new features that we find very handy in instructional use. This manual assumes that you are using System 7 and *Mathematica* 2.2. Contact your local authorized Apple reseller or your Apple campus representative for more information on getting a copy of System 7. If you wish to use System 6 (and *Mathematica* 2.0 or 2.1), then you should contact your resident Mac guru for help on installing the fonts and other resources.

If you're ready, get going!

Understanding the Macintosh desktop

The Macintosh interface is based on a desktop metaphor. A person interacts with the computer the same way he or she interacts with a desk—using tools, folders, files on a desk, as well as a trash can for discarding work.

The biggest part of your interaction with the Macintosh (outside of *Mathematica*, of course) will be with the **Finder**. Every Macintosh automatically starts the **Finder** when it is powered up. The usual appearance of the desktop after start-up is like this (not including the explanations, of course):

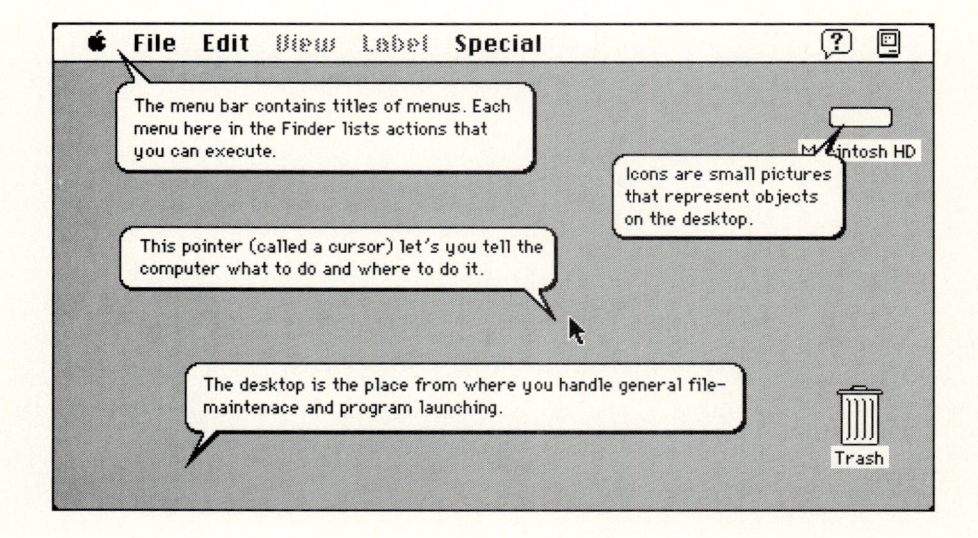

Know your icons

As mentioned earlier, the Macintosh uses the desktop metaphor to make using a computer similar to that of managing files on your desk. Icons are the little pictures that the Macintosh uses to represent objects in the real world. Here are a few of the icons that the Macintosh uses:

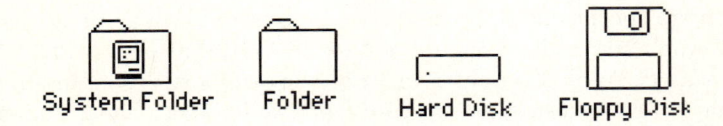

These icons can be thought of as file cabinets or drawers. They hold both the tools you use to create your work (application programs) and the work itself (documents). Folders can contain applications, documents, or other folders. They allow for easy organization of your work. The **System Folder** is special: It contains the information necessary to start up your Macintosh.

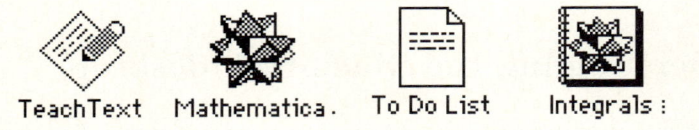

Applications are programs, computer instructions. The **TeachText** application allows you to compose, save, and print short letters. **Mathematica** allows a user to do numeric, symbolic, and graphic computations. Documents are the files you create with applications. Here we see a **TeachText** document (which is a list of things to do) and a **Mathematica** document (which is about integrals).

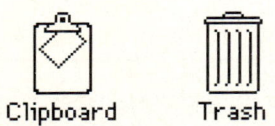

Some icons have very unique functions. They correspondingly have unique shapes. Here we see **Clipboard** (temporary storage of information) and **Trash** (where to throw away other icons).

Mousing around

It usually takes a little practice to get comfortable with the mouse, so don't be discouraged. There are five standard mouse techniques that will let you do all of your work on the Macintosh (except typing text).

- Point:

You *point* to an object on the desktop by moving the mouse until the tip of the arrow is on top of the object. In the figure above, the arrow is pointing to the hard disk icon (which is named **Macintosh HD**).

- Click:

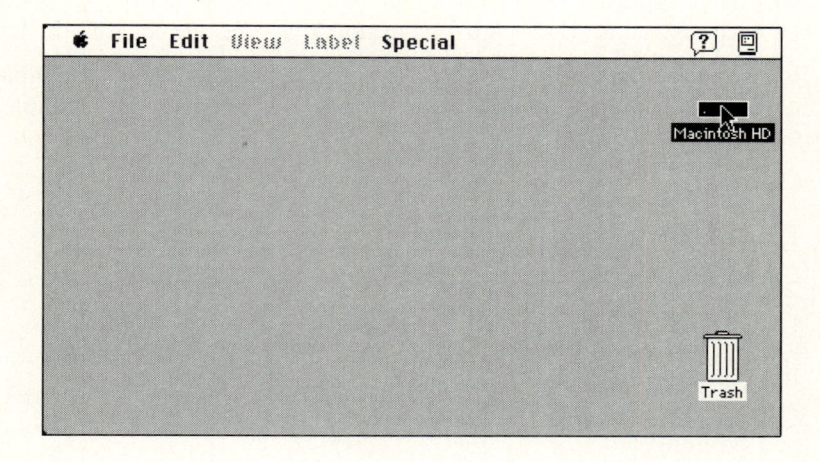

You *click* by first pointing to an object, then pressing and releasing the mouse button once. When an icon is clicked, it becomes *highlighted* or *selected*. The highlighted icon is black (if you have a color screen, it will be dark colors). A highlighted icon means that the icon is ready to be acted on. In the figure above, the hard disk icon called **Macintosh HD** is highlighted.

- Press:

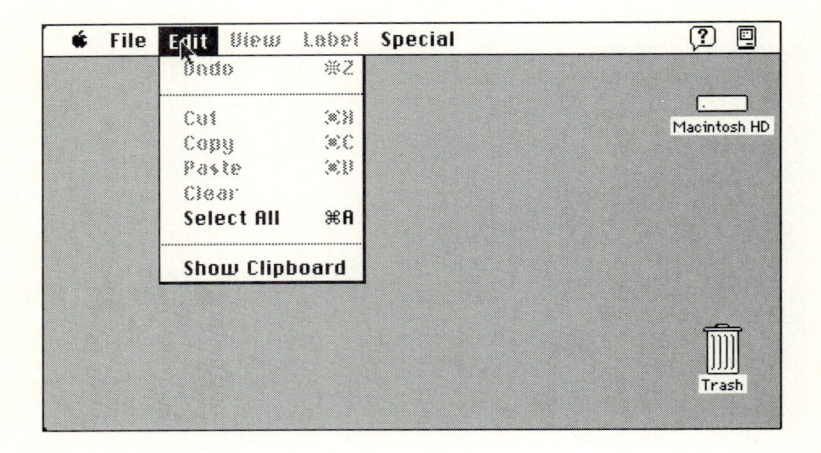

You *press* by first pointing to an object, then pressing and holding the mouse button. You can press all of the titles in the menu of the **Finder** (or Apple, **File**, **Edit**, etc.). You will see all of the selections for each of the titles.

This is how a command is issued in the Macintosh. First, you select an object (a hard disk, application, document, or folder) then an action. It is analogous to a sentence:first noun, then verb. All available actions are located in the menus.

In the figure above, the **Edit** menu is pressed to see the actions available underneath it. Note that some of the options are grayed out, since in this example, an icon was not selected beforehand.

- Drag:

You *drag* an object by first pointing to it, then pressing it, then moving the mouse. Here you can think of the pointer as an extension of your hand. Clicking and holding the mouse button is like grabbing an object, and moving the mouse is like moving the object around the desktop.

In the figure above, the hard disk icon is being dragged. When the mouse button is released, the hard disk icon will move from its old location to the location of its outline. This technique can also be used to move other icons from one location to another. For example, if you drag a folder or a document onto the **Trash** icon, the folder or document will be put into the **Trash**. You could then select **Empty Trash** from the **Special** menu to discard a file permanently.

- Double-click:

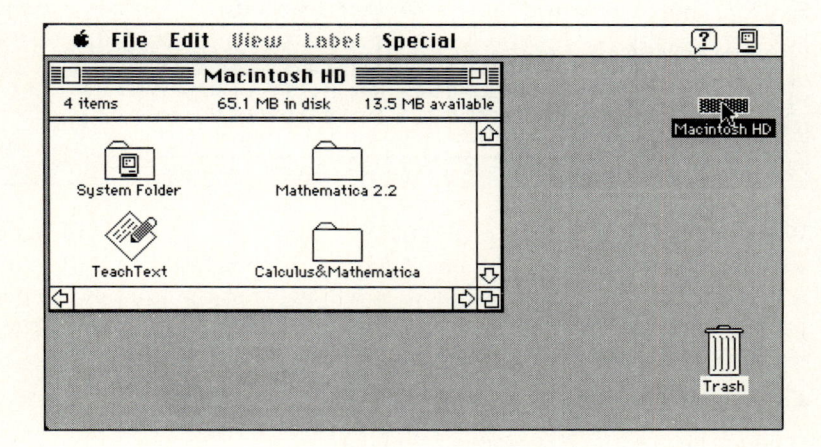

You *double-click* by first pointing to an object, then clicking twice without moving the mouse. The timing sometimes takes a little practice. Double-clicking is a shortcut for an action that is available somewhere in the menus. Double-clicking an icon on the desktop is a shortcut to selecting the icon, then choosing **Open** from the **File** menu. In this example, the hard disk icon was double-clicked to show you its contents, which are displayed in a window.

You don't have to remember all of the shortcuts. As long as you remember the two-step process (first select an object, then select an action), you will be able to execute all of the Macintosh commands.

Understanding the windows

All applications in the Macintosh environment use windows to show you the contents of an object. In the **Finder**, windows tell you what's inside disks and folders. Applications will open windows to documents. Let's first look at a typical **Finder** window again.

20 Calculus&*Mathematica* with Macintosh

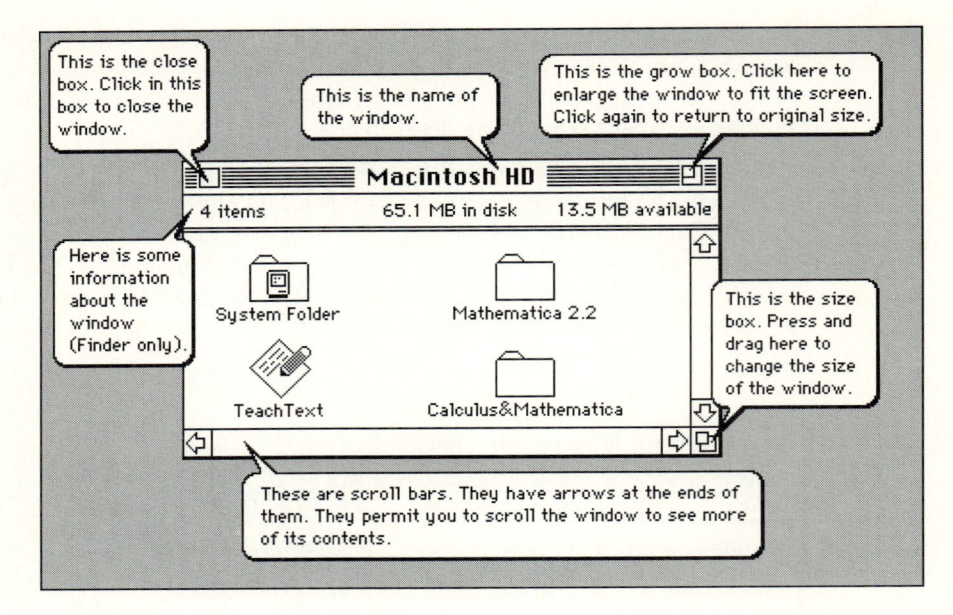

[Illustration: Macintosh HD Finder window with callouts labeling the close box, window name, grow box, information area (Finder only), size box, and scroll bars. Window contains: System Folder, Mathematica 2.2, TeachText, Calculus&Mathematica. 4 items, 65.1 MB in disk, 13.5 MB available.]

Windows can also overlap one another. Think of it as a pile of papers. You can shuffle the order to see which one is on top. If you click on a part of a window that is underneath your current window, it will move to the front of the pile. This gives you a lot of flexibility to look at information. Here is a look at the overlap and scrolling of windows:

• Overlapping windows:

Double-click on the **Mathematica 2.2** folder.

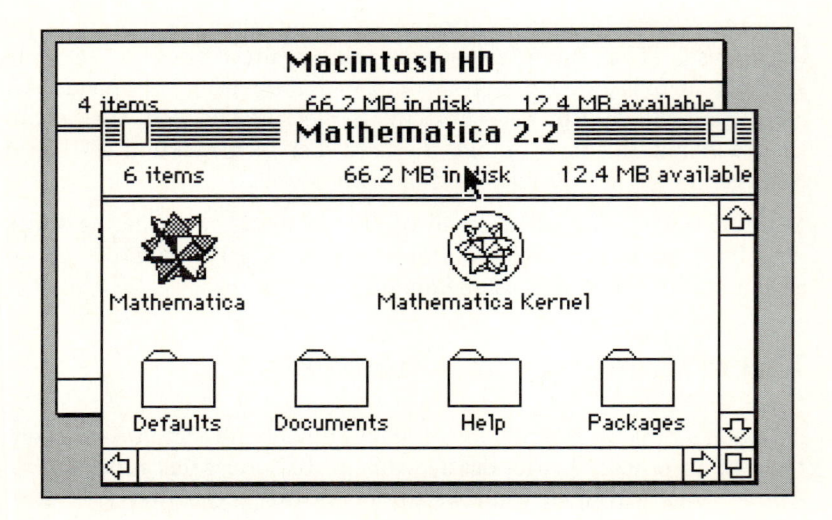

[Illustration: Macintosh HD window (4 items, 66.2 MB in disk, 12.4 MB available) with Mathematica 2.2 window overlapping in front. Mathematica 2.2 window: 6 items, 66.2 MB in disk, 12.4 MB available. Contents: Mathematica, Mathematica Kernel, Defaults, Documents, Help, Packages.]

Calculus&*Mathematica* with Macintosh   21

The **Mathematica 2.2** window opened up. It is overlapping the old **Macintosh HD** window. To bring the rear window to the front, just click on any part of the rear window.

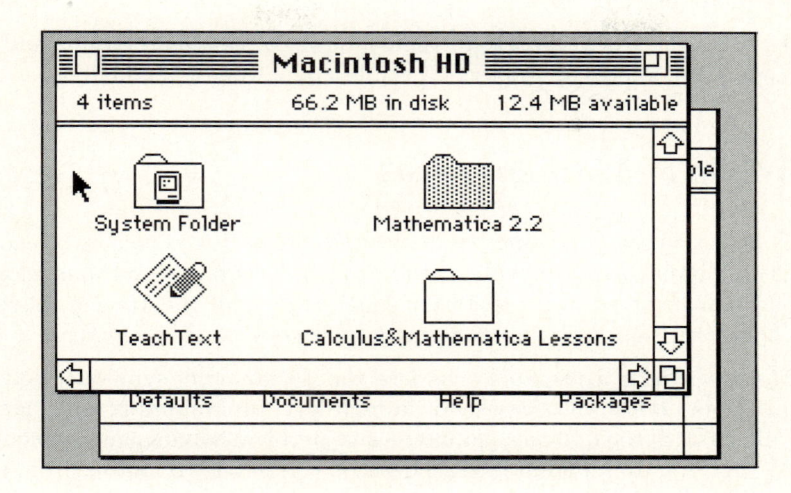

The left part of the rear window has been clicked. It has moved in front of the old window. Note how the **Mathematica 2.2** icon is grayed out. This means the folder is open.

- Scrolling in windows:

Double click on the **Packages** folder in the **Mathematica** window. Here's what you get:

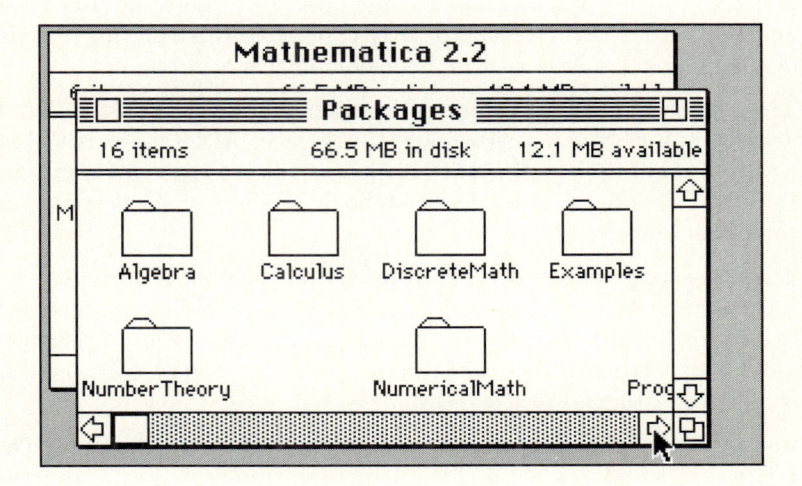

Notice the start of a name at the lower right and the grayed horizontal scroll bar (with the scroll box between the two arrows). This means you can scroll right to see more information.

As the right arrow is pressed, the contents of the window shift, and you see more icons. You could also grab the scroll box and drag it manually.

What is *Mathematica*?

This section is an introduction to using *Mathematica*. If you are already comfortable with *Mathematica*, you will want to skip this section. If you want more information on the features of your versions of *Mathematica* for the Macintosh, read the *User's Guide for Macintosh* that came with your copy of *Mathematica*.

Mathematica is a program that lets you do numeric, symbolic, and graphic computations. It can also be used as a high-level programming language. It is used in Calculus&*Mathematica* as an interactive problem-solving environment. And it runs on anything from mainframes to desktop computers to laptops from many different computer companies.

There are two parts to every version of *Mathematica*: the Kernel and the Front End. The Kernel is the brain; all of the computations are done with the Kernel. Since the Kernel is the same on every platform, *Mathematica* instructions are easily ported from machine to machine, and the output is exactly the same.

The Front End is how users communicate to the Kernel. The Front End differs from each different platform. For most mainframe (UNIX-type) machines, the Front End is nothing more than a command line interface, where the user types a command and the output is then displayed. It is possible to string together lengthy groups of commands, but it is very cumbersome in this environment.

The advanced Front Ends (currently shipping for Macintosh, NeXTStep, and Windows computers) have a Notebook^{TM} interface. A Notebook is a file that organizes text and graphics in *cells*, much like a word processor's different paragraph styles. But the cells are what set the Notebook versions of *Mathematica* apart from its command line cousins.

There are many different types of cells, and each has a particular function. Input cells contain *Mathematica* code and can be *executed* to produce output cells (for textual or numeric output) or graphic cells (for plots and other pictures). There are different types of text cells to put comments around *Mathematica* code, as well as different title and section cells.

Cells can have different formats, such as font, font size, color, etc. They can also be grouped together in a hierarchical structure, similar to a word processor's outlining capability. This makes it easy to organize related concepts, which is very helpful for an interactive course.

The portability of a *Mathematica* Notebook is one of its best features. It allows users to transfer a Notebook from a Macintosh to a Windows to a NeXTStep computer without losing any information or formatting. And since the Notebooks are text files, it is even possible to send them via electronic mail.

Let's see how to get started in *Mathematica*!

Composing a simple *Mathematica* document

First, locate your copy of *Mathematica* on your hard disk.

Double-click on the **Mathematica** icon to start *Mathematica*. To show you the menu bar in full size, the next screen shown has the right side cropped off. *Mathematica* automatically opened a new window for you. Notice that the **Mathematica** window has many similarities to **Finder** windows; it has a close and grow box as well as scroll bars. But it also has a few differences. The lower left of the window contains some extra information.

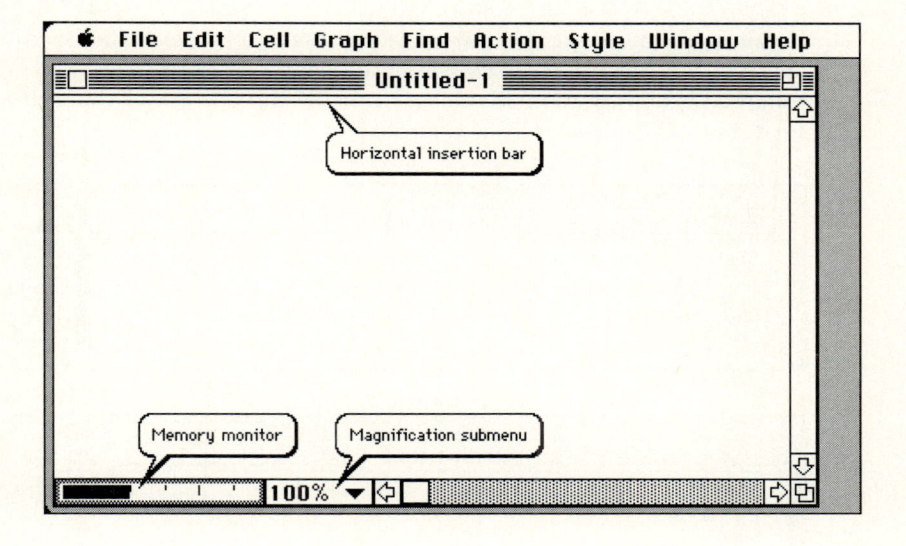

The *horizontal insertion bar* tells you where the next input (via keyboard or mouse) will be. Since there is nothing yet in your Notebook, the insertion bar is at the top of the window. The *memory monitor* is a scale that tells you how much memory is left for *Mathematica* Front End. When it fills up, *Mathematica* will become slow and unresponsive. The *magnification submenu* lets you change the zoom factor on a document. You can set it between 50% and 300%.

To insert text into your document, just start typing (when there is a horizontal insertion bar). By default, *Mathematica* will put the characters into an *input cell*, and the characters will be in the style `Courier bold`.

24 **Calculus&*Mathematica* with Macintosh**

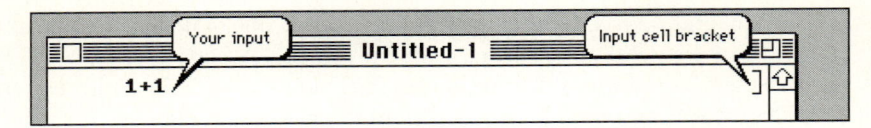

[Figure: Untitled-1 window showing "1+1" with labels "Your input" and "Input cell bracket"]

While the cursor is still in the cell, press the **Enter** key or **Shift-Return** keys simultaneously to execute the input cell and generate an output cell.

[Figure: Untitled-1 window showing In[1]:= 1+1, Out[1]= 2, with labels "Your output", "Input/output group bracket", and "Output cell bracket"]

By default, *Mathematica* automatically grouped the input and output cells together. Also, `In[1]:=` and `Out[1]=` were automatically attached to the input and output cells. These labels just tell you how many times you've executed any input cell.

If you double-click on the outermost cell (the group cell), the group collapses and only the top cell will show, which is the input cell in this case. You can identify a collapsed group by the little down arrow at the bottom of the cell, as well as the small box to the lower left of the cell.

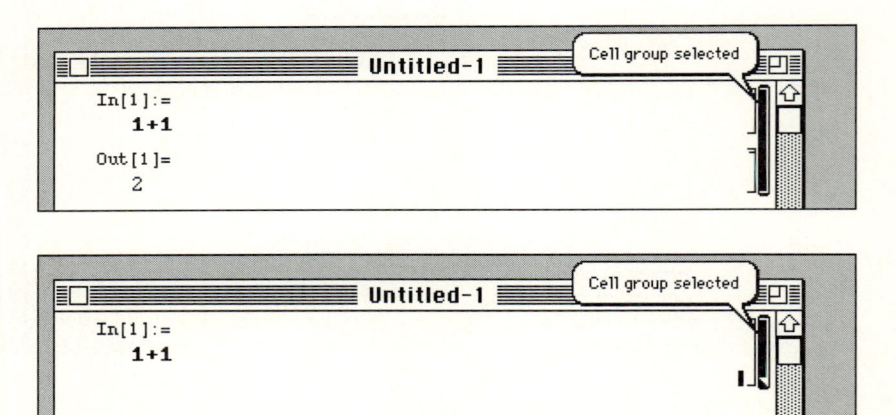

[Figure: Untitled-1 window showing In[1]:= 1+1, Out[1]= 2, with label "Cell group selected"]

[Figure: Untitled-1 window showing In[1]:= 1+1 (collapsed), with label "Cell group selected"]

Groups can contain cells or other groups. Thus, cells can be layered as deeply as you want. You can create your own groups by selecting a range of cells and selecting **Group Cells** from the **Cell** menu. Of course, the top cell will be the only one visible when you close the group.

Now put text comments above and below the input/output cell group, then group them all. First, move the cursor between the title bar of the window and the first cell. The cursor will then turn into a horizontal I-beam. Click once to create a horizontal insertion bar. Start typing some text.

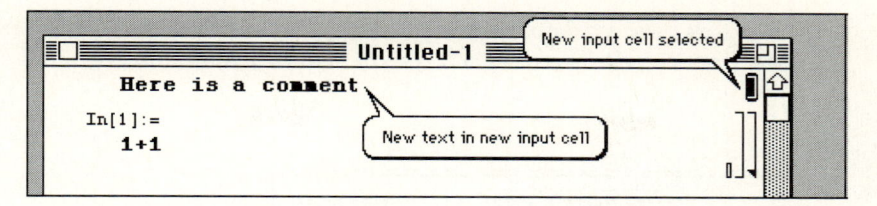

```
┌─────────────────────────────────┐
│ ▪️════════ Untitled-1 ════════ New input cell selected
│    Here is a comment
│ In[1]:=            New text in new input cell
│    1+1
└─────────────────────────────────┘
```

Now select the new input cell, and then select **Cell Style** from the **Style** menu. You will see a cascading submenu. Then select **Text** (or use the keyboard shortcut **Command-7**).

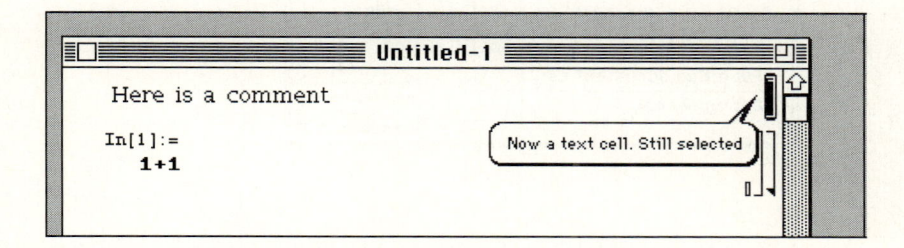

```
┌─────────────────────────────────┐
│ ▪️════════ Untitled-1 ════════
│    Here is a comment
│ In[1]:=            Now a text cell. Still selected
│    1+1
└─────────────────────────────────┘
```

See how the font changed to the Text font (**Calculus** in this example).

Now add another comment underneath the input/output group, and make that into a text cell.

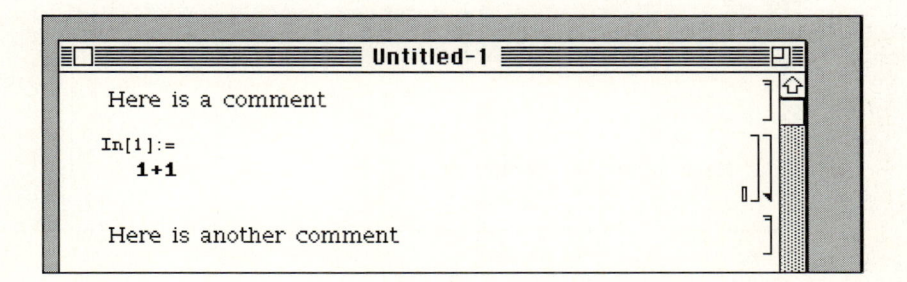

```
┌─────────────────────────────────┐
│ ▪️════════ Untitled-1 ════════
│    Here is a comment
│ In[1]:=
│    1+1
│
│    Here is another comment
└─────────────────────────────────┘
```

You can group these cells together:

Begin by clicking on the uppermost cell (the first Text cell), but don't release the mouse button. Drag the mouse down until all of the cells are selected. Then select **Group Cells** from the **Cell** menu.

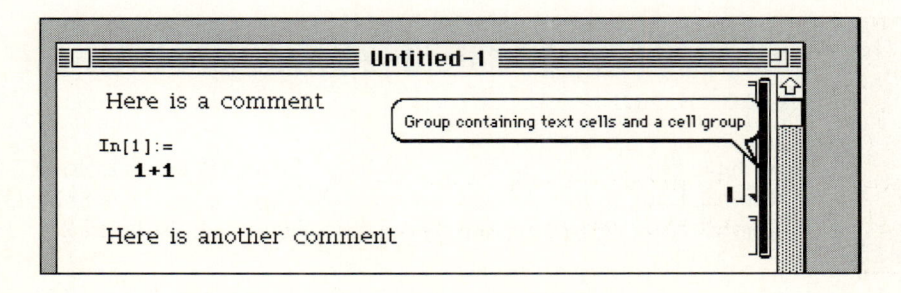

```
┌─────────────────────────────────┐
│ ▪️════════ Untitled-1 ════════
│    Here is a comment
│                    Group containing text cells and a cell group
│ In[1]:=
│    1+1
│
│    Here is another comment
└─────────────────────────────────┘
```

Now you have a group inside a group! Save your Notebook to the hard disk. Select **Save** from the **File** menu. Then type the name of your Notebook and click on the **Save** button.

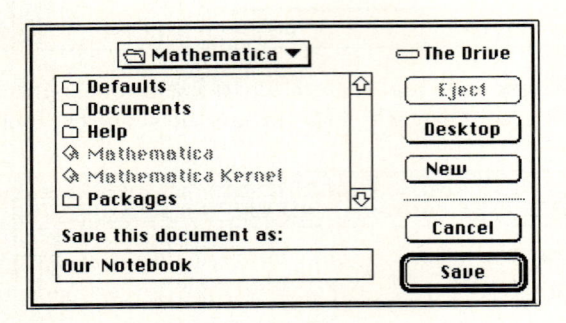

Note how the **Open/Save** file dialog box of *Mathematica* is the same as other programs. Now you can open the Notebook by selecting **Open** from the **File** menu to check your work.

Note how the title bar now has the title of the Notebook as you typed it in (namely, **Our Notebook**). Also, the **In[1]:=** label is gone, since *Mathematica* doesn't know that this is the same Notebook that you started with.

You have just completed your first Notebook!

Installation of Calculus&*Mathematica* for the Macintosh

You have two installation disks: **CM_Install_1** and **CM_Install_2**. First, insert the disk **CM_Install_1** and follow the instructions below. After you have finished with the disk **CM_Install_1**, repeat the procedure for the second disk **CM_Install_2**.

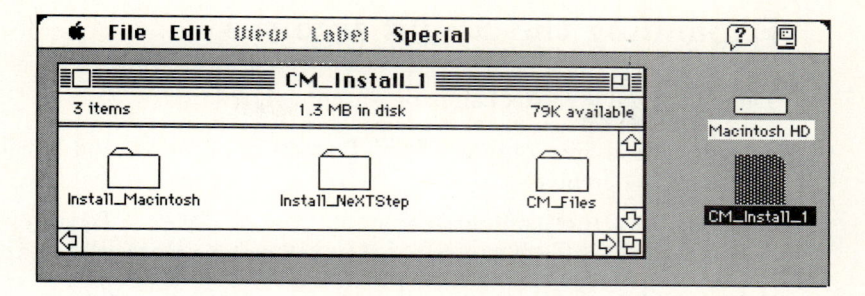

You will see icons of folders called **Install_Macintosh**, **Install_NeXTStep**, and **CM_Files**. Drag the folders **Install_Macintosh** and **CM_Files** onto the hard disk icon to copy them. (The **Install_NeXTStep** folder is for installing Calculus&*Mathematica* on a NeXTStep computer.)

If you have purchased the full version of Calculus&*Mathematica* (all four books), the other disk will have a folder called **More_CM_Files**. You should also drag this folder to your hard disk. Now let's see what's in these folders.

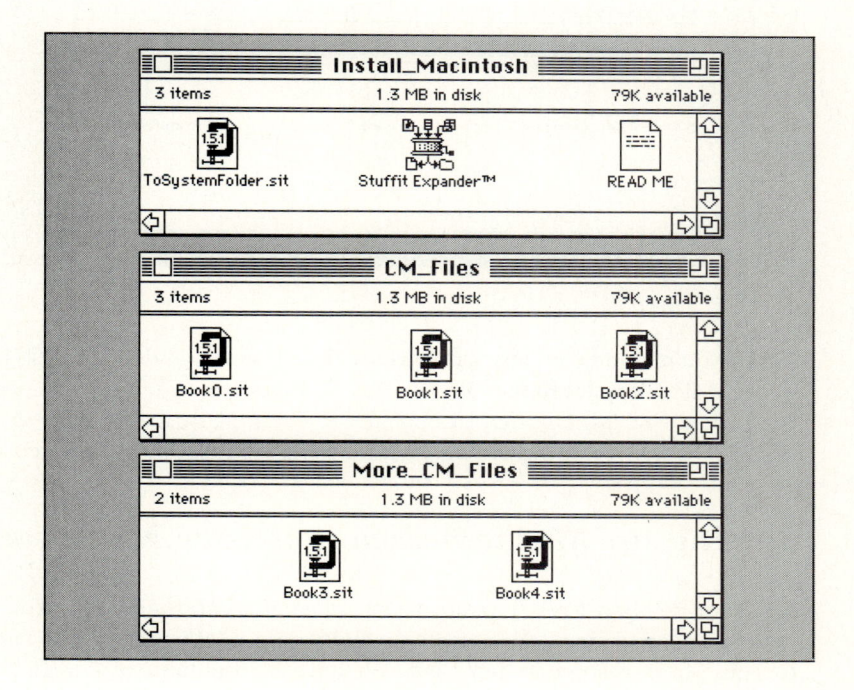

The **CM_Files** and **More_CM_Files** folders contain the Calculus&*Mathematica* lessons in compressed documents called Stuffit^{TM} archives. The **Install_Macintosh** folder contains an application, **Stuffit Expander**^{TM}, that expands **Stuffit** archives. If the lessons were distributed without compression, it would take over four floppy disks!

Expanding the *Stuffit* archives

There are two ways you can expand the archives:

- Drag each archive onto the **Stuffit Expander** application, and it will automatically expand the archive.
- Double-click the **Stuffit Expander** application, and choose **Expand** from the **File** menu. From the resulting dialog box, select the archive you wish to expand.

Stuffit Expander will put its output in the same folder the **.sit** file lives in, so you need to drag the **.sit** files into your new **Calculus&Mathematica** folder before you unstuff them if you are going to use that utility.

Book0.sit, **Book1.sit**, **Book2.sit**, **Book3.sit**, and **Book4.sit** expand into **0.FeelofMathematica**, **1.Growth**, **2.Accumulation**, **3.2Dand3DMeasurements**, and **4.Approximation** folders, respectively. Put all of these folders into a new folder called **Calculus&Mathematica**.

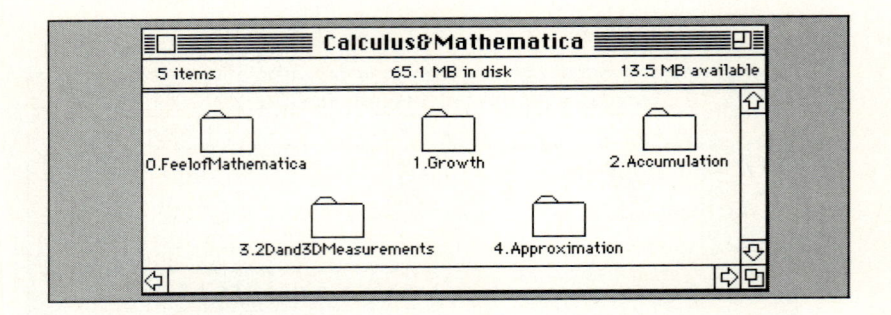

When you expand the **ToSystemFolder.sit** archive (the **.sit** stands for **Stuffit**), you will get a folder called **ToSystemFolder**. This folder contains the Calculus font necessary to use the Calculus&*Mathematica* lessons. Drag the font suitcases onto the **System Folder**, and the **Finder** will automatically install the Calculus font into its correct place.

Setting up *Mathematica* and your Macintosh

Now, you're almost ready to start using Calculus&*Mathematica*. You need to check just one more thing. You need to make sure *Mathematica* has enough memory to run. It is recommended that you have a minimum of 8 MB (megabytes) of RAM (random access memory, the physical electronic memory) installed on your computer. Unless you have much more than 8 MB, you will probably need to use virtual memory to get acceptable performance.

What is virtual memory? Well, your Macintosh has only so much RAM, which is the memory where programs are loaded when launched from the **Finder**. Virtual memory is a feature of System 7 that allows a portion of your hard disk to act as physical memory. Thus, you can run more programs or, for our case, one program a lot more efficiently.

Not all Macintosh computers are virtual-memory-capable. To check if your machine is, open the **Memory** control panel.

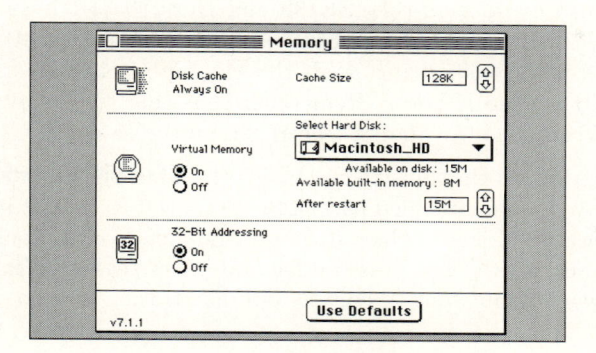

If your Macintosh is virtual-memory-capable, you will see a **Virtual Memory** setting in the **Memory** control panel. Set this to **On**, then use the arrows to up the memory.

The most recent Macintoshes (1989 or so) have a setting in the **Memory** control called **32-Bit Addressing**. This will allow you to access more than 13 MB of memory, virtual or physical. If your Macintosh allows this to be set, turn it on.

Now that you've set the **Memory** control panel, you must prepare *Mathematica* to use this extra memory. Go to the **Mathematica 2.2** folder, and open it up. Select the **Mathematica** icon and select **Get Info** from the **File** menu. If you are using *Mathematica* 2.2 or later, you will need to **Get Info** on the **Mathematica Kernel** as well. The two windows you will get are as follows:

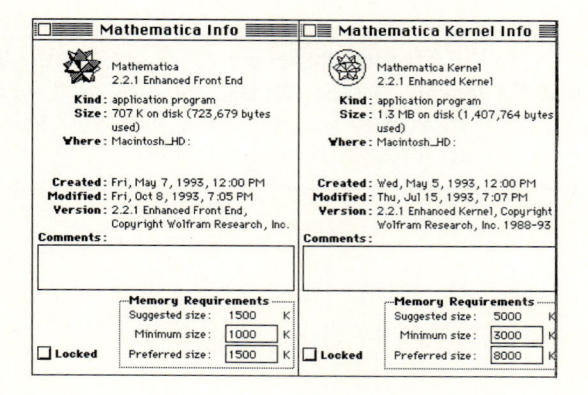

To set the Front End and the Kernel to get more memory, just place the cursor into the **Preferred Size** field, click, and type in the number of kilobytes you want the application to use. Example: Say you set the virtual memory to 20 MB. The **Finder** and **System Software** will take approximately 3 MB. Set aside 5 MB just in case you run other applications concurrently with *Mathematica*. So we have 15 MB left for *Mathematica*. At most, the Front End should have two-thirds of the memory you want to set aside. So give the Front End 5 MB, which leaves 10 MB for the Kernel.

For older Macintosh computers (like SE/30 and IIcx) the setting is at 20 MB of virtual memory (5 MB for the Front End, 15 MB for the Kernel). On the new machines (Centris 650), the setting is all the way up to 40 MB (15 MB for the Front End, 25 MB for the Kernel). Remember that this will take up equal space on your hard disk, so be sure you don't need the hard disk space.

You have one more important thing to do. Inside the **0.FeelofMathematica** folder, you'll find a *Mathematica* document called **CalcMath.ma**. You need to drag it into the **Defaults** folder in the **Mathematica** folder. Then start up *Mathematica* by double-clicking on its icon. Once it has started up, press on **Edit** and select **Preferences/File Locations**. You'll get a dialog box which you want to edit to look like this.

Click on **OK**. You are ready to go.

Now go have fun. Go to the **0.FeelofMathematica** folder and read the **Greeting**. Then start the **FeelofMathematica** lessons, **Numbers**, **Algebra**, and **Plots**, and work your way through them doing everything they suggest. After that, you'll be comfortable with *Mathematica*, and Calculus&*Mathematica* lessons.

Calculus&*Mathematica* with NeXTStep

**by David Wiltz,
Calculus&*Mathematica* Development Team**

Welcome to Calculus&*Mathematica* with NeXTStep. By deciding to explore mathematics with the aid of the Calculus&*Mathematica* software, you have taken the first step on what will hopefully be a deep and meaningful understanding of what mathematics is and what it can do for you. The NeXTStep section of this guide has been designed to give a basic familiarity with the NeXTStep operating system and *Mathematica* with NeXTStep, as well as to provide technical assistance with the installation and use of the courseware. Since it is written for many levels of users, you may wish to skip around to find what sections are applicable to your needs. While this section of the guide was written on and for actual NeXT hardware, most of the information is either directly or closely related to the NeXTStep operating system on Intel and other platforms. Since the best way to learn about a computer is to sit down and get to work, find a NeXT, make yourself comfortable, and begin.

NeXTStep Basics

Unlike the Macintosh or PC platforms, NeXTStep requires each user to have a login name and password, which it uses to keep private files private, among other things. When you first sit down at the computer, you'll be presented with a login screen which will look something like this:

32 Calculus&*Mathematica* with NeXTStep

In the login field, please enter your login name, which was given to you by the administrator of the computer. Then press the **Enter** key, enter your password, and press the **Enter** key again. In a few moments, the display will change from the login screen to what is called the **Workspace**, akin to the **Desktop** on Macintosh or Microsoft Windows computers. If your account is a new one, the **Workspace** will probably look a bit like the one shown below. **Mathuser** will be replaced by your login name, but the rest should look familiar.

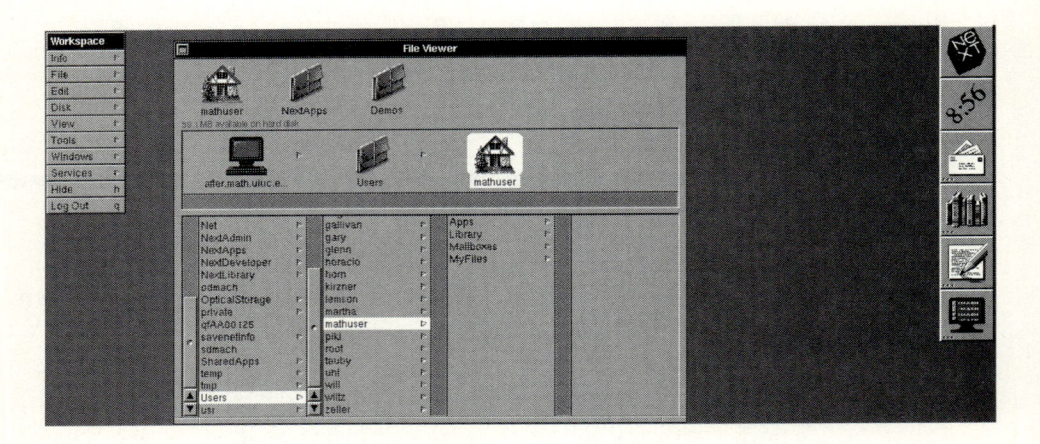

This is a good time to go over some of the underlying features of the NeXTStep operating system, as well as the hardware required to support many of the operations that make NeXTStep unique. Whether you are on an actual NeXT computer, an Intel PC running NeXTStep 486, or some other flavor of computer running a native version of NeXTStep, you will find that the accompanying figures look much like what you have on your screen. This is due to the distinctive look of the NeXTStep graphical interface. This interface allows you to use the mouse to move a pointer (it looks like an arrow, usually) and click on the operation you want to perform. Those familiar with the Macintosh or Microsoft Windows interfaces should be right at home in the NeXTStep operating system, and those who are new to graphical interfaces should find little trouble learning how to navigate in the graphical environment. In fact, the graphical interfaces on all of these platforms are so similar that, if you learn about clicking, dragging, and file management on one, you surely will have a basic understanding of what's up on the others as well. Check this out. If you would like to see some of this explained a bit differently, peruse the sections on starting out with the Macintosh or Microsoft Windows.

As mentioned earlier, many of the operations used in NeXTStep are driven by the use of the mouse, which is why NeXTStep is often referred to as a point-and-click interface. You use the mouse to move the pointer to the operation you wish to perform, and click on the mouse button (usually the left one) to cause the operation to happen. You can also use a mouse to perform click-and-drag operations, which are used to highlight sections of text or move files between directories (folders).

To click and drag, move the mouse to the area you wish to drag from (either the beginning of the text to highlight or the name of the file you want to move), and click the mouse button, BUT DO NOT RELEASE THE MOUSE BUTTON. Now, while still holding the button down, drag the mouse to the final destination (the end of the text to highlight or the destination of the file). Now you can release the mouse button, which finishes the operation. One other important mouse operation is double-clicking. This is closely related (as the name suggests) to clicking, but requires that you press the mouse button twice in rapid succession. This is used to open applications from the dock (explained below) or to open files from the browser (also explained below).

The NeXTStep **Workspace** has three primary areas, the **Application Dock**, the **Menu Area**, and the **File Viewer**, as seen in the above figure as the right, left, and center areas of the screen. Located in the **Application Dock** are buttons (called icons) which allow you to access commonly used programs quickly. You launch these programs by moving the mouse so that the arrow is over the application you wish to use and double-clicking on the icon. Some common icons, and the applications they represent, are shown below.

Mathematica   NeXTMail   Edit   Terminal   Recycler

You can add icons to the **Application Dock** by dragging their icon to the dock, then releasing them. You will probably want to do this with *Mathematica*, since you will be using it a lot in the near future. On the left side of the screen there is a menu of operations that can be performed. Clicking on any of the menu items with an arrow next to it will cause another menu to appear. Clicking on a menu item with no arrow causes that operation to be performed. Items with letters next to them may also be activated by holding down the **Command** key (next to the spacebar) while pressing the letter. Each application has its own set of menus, which contain commands necessary to the application. You should take some time when you first start an application to familiarize yourself with what operations are available, as well as how to get to them.

In the center of the **Workspace** (sometimes called the **Desktop**) is a window called the **File Viewer**. This is where all the files, folders, and applications (which are really only special types of files) are presented. The top portion of the **File Viewer** is called the shelf and is a place where you can keep a list of commonly used files or folders accessible at all times. Clicking on the icon that looks like a house, for instance, will always get you back to your home directory. You can add folders to this area by dragging them to the place you wish to put them. This does not move the folder but creates a reference (called an *alias* on the Macintosh) to the folder's location. The area directly below the shelf contains a list of the current hierarchy of folders you have entered to get to where you are. This is also the place where you can click on file and folder names to change them. Immediately below this area is the file browser section of the **File Viewer**. It shows a list of all the available files and folders on the computer. Clicking on any of them will move you to that directory (folder), which changes how the area below the shelf looks. The different ways that the **File Viewer** can look are determined by the menu selections under view. The style shown in this guide is called *browser*, which shows the most simultaneous information of all three possible views.

One more basic item of the NeXTStep interface is how information is presented in windows. The following figure is representative of a generic NeXTStep window.

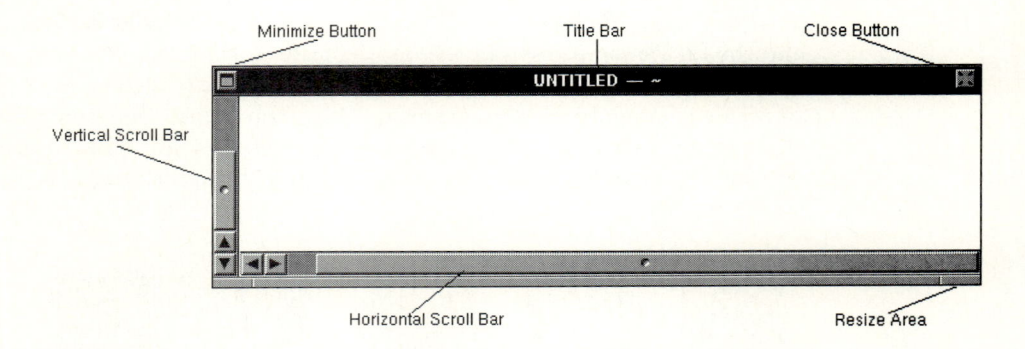

Following is a list of the areas of the window and what can be done with them:

Close Button: This is a solid X if the document has not changed since the last time it was saved, a broken X if it has. Clicking on this button will close the window. If the information has not been saved, you will be asked if you really want to do this.

Minimize Button: This button shrinks the window and places it along the bottom of the screen. This is useful to help remove some unnecessary clutter from the screen. Double-click on the minimized icon to resize the window.

Title Bar: This shows the name of the window. Clicking and dragging within the title bar will move the window around on the screen.

Vertical Scroll Bar: This slider moves the viewable area of the window up and down. This will not appear until you have too much information to be displayed in the window by itself. Click and drag the scroll bar to move the displayed information, or click on the up and down arrows.

Horizontal Scroll Bar: This works like the vertical scroll bar but controls left and right shifts of the display.

Resize Area: Click in this area and drag to resize the window.

Now that you have a basic understanding of the NeXTStep interface, we can explore how to use this knowledge to do things with NeXTStep. Take mouse in hand, and let's go.

One important menu item creates new folders. You will create folders (also called directories) to hold your personal homework and other files you will create. Make a new folder now by clicking on the **File** menu item. When you do so, the menu will change to look like the following:

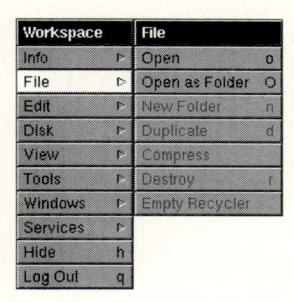

Workspace	File
Info ▷	Open o
File ▷	Open as Folder O
Edit ▷	New Folder n
Disk ▷	Duplicate d
View ▷	Compress
Tools ▷	Destroy
Windows ▷	Empty Recycler
Services ▷	
Hide h	
Log Out q	

Now click on the **New Folder** menu item, and a folder called **NewFolder** will appear in the **File Viewer** window. Notice that you can also create new folders by typing **Command-n**. (There is an **n** next to **New Folder** in the menu.)

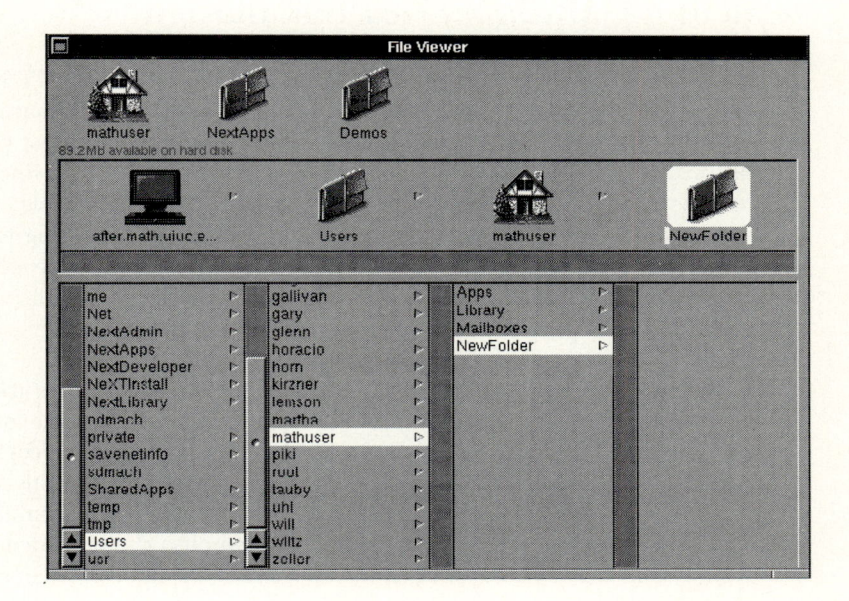

You should give the folder a name that means something to you. While the name **NewFolder** is still highlighted, type in a name that is related to the types of files you will be storing in this folder. If you wish to store Calculus&*Mathematica* homework in there, you might want to name it **C&M Homework**. Note that, unlike DOS, you can use spaces in file and folder names and as many characters as you wish. You may want to be careful with this type of naming scheme, however, especially if you will be moving files back and forth between different machines. See the section on file transfer for more information on this.

You may have noticed that some menu items are dark, while others are gray. The gray items are commands that you cannot perform, usually because they make no sense in relation to what you are doing at the time.

To change the name of a folder or file at a later time, you need to click once on the file, which will place it on the second line of the **File Viewer**. Now click on the name under the highlighted file. You should get a flashing bar (a cursor) inside the name. You can now modify the name from the keyboard as before.

To move a file, click on the file and, while holding the mouse button down, drag its icon to the new location. The same operation will copy files from floppy disks to a new directory.

To delete a file, click on its name to select it, and in the **Workspace File** menu, select **Destroy**. Now, from the **File** menu, choose **Empty Recycler**.

Now that you have some basic familiarity with some NeXTStep file operations and with the system in general, it is time to install the Calculus&*Mathematica* software.

Calculus&*Mathematica* Installation

There are two ways to install Calculus&*Mathematica* under NeXTStep. The suggested way is to install it so that everyone who uses the computer can use the specialized font and keymapping, as well as the optimal defaults. For the full installation, you need to be able to log in as the superuser, *root*, since some of the commands require modifying NeXT font libraries and other things only root can do. If you don't have root privileges on the machine you are going to use, you can still install the course software. However, you will be the only one who will be able to access the fonts, keymaps, and defaults, since you won't be able to put them in the right places on the machine.

The install process: The NeXTStep release of the Calculus&*Mathematica* courseware is included as part of the Macintosh distribution. In order to install the courseware and supporting materials into the necessary directories, a UNIX Makefile has been included with the Macintosh source. Executing the `make` command, explained below, will cause the Calculus font, a special Calculus keymapping, and all Notebooks that are present to be installed onto the hard drive. You should read the accompanying `NextREADME` file, located in the NeXTInstall folder on the installation disk, which documents the details of the `Makefile`, as well as how to make any necessary changes.

Which files are needed: Prior to installation, you need to create a directory (folder) that will eventually contain the courseware. Choose where you want to install the lessons and create a folder there. Remember that this can be done from the **Workspace** by pressing **Command-n**, then renaming the folder. Once this directory exists, copy the following files from the Macintosh distribution disks to this newly created directory. (Note to those who wish to use a network to transfer files: If using **ftp** to move the files from one computer to another, make sure that all files are moved in binary mode. If not, the installation will fail.)

> `Makefile`: Contains the commands necessary for the installation of Calculus&*Mathematica*.
>
> `NextREADME`: Installation notes and instructions for modifying the `Makefile`.
>
> `NextSpecific.tar.Z`: A file archive that contains the same installation utilities, the Calculus keymapping, the Calculus postscript fonts, and new *Mathematica* default Notebooks that have been set for ease of use with Calculus&*Mathematica*.
>
> `bookx.sit`: A file archive that contains the actual Calculus&*Mathematica* lessons. The `x` in the file name stands for a particular number. Please copy all of the books that you wish to install. Note that books 3 and 4 are located on the second distribution disk, **CM_Install_2**. For instance, copy `book1.sit` and `book2.sit` if the lessons that accompany books one and two need to be installed. The `make` utility will check for any available book files during installation and extract them to the correct place. This place will be a directory within the one you created.

After all the files have been copied, open a **Terminal** application window by double-clicking on **/NextApps/Terminal.app** or the **Terminal** icon on the right side of the **Workspace**. In the **Terminal** window, move to the directory you just created. (If you are root user of the computer, you know how. If you aren't, type the following line at the prompt):

`% cd ~/<place the name of your folder here>`

Now type the following command:

`% ls`

You should see the names of the files you just copied listed out for you.

Now execute one of the following commands:

If you are the superuser,

`% make global`

If not,

`% make personal`

(Some warning messages regarding duplicate fonts may appear during installation; this is normal.)

The `make` utility will read the `Makefile` and execute all commands within it. All `bookx.sit` files will be expanded, with the Calculus&*Mathematica* Notebooks within them stored in the correct directories. The Calculus font and keymapping will be moved to the correct places on the system, and the font directory will be rebuilt to allow the Calculus font to be used.

Once the execution is complete, enter the `ls` command again. You will see the lesson folders and some remnants of the installation procedure. You can remove the extra stuff if you wish.

Provided that the `make` utility executes with no errors, Calculus&*Mathematica* should now be ready to run. Before continuing, please check that the fonts and keymapping were correctly installed. To check the fonts, while still in the **Terminal** application, press **Command-t**. This will display a font panel that looks something like the following:

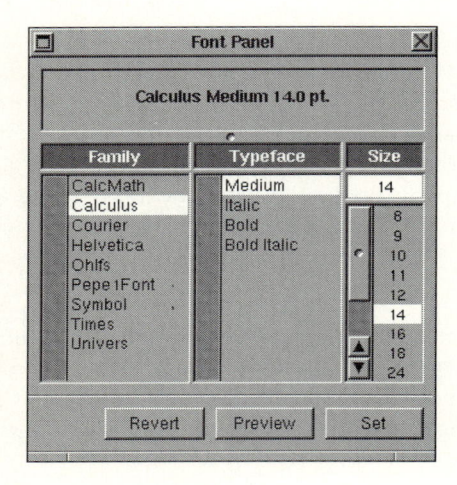

If you do not see the **Calculus** font, check the font panel in another application (*Mathematica* is a good choice here) and check if it shows up now. If so, everything is fine; if not, check the **Terminal** window for install errors.

To check if the keymapping is in the system, go to the **Preferences** application and check international settings (the button with all the flags). The **Calculus** keymapping should be visible under the **Keyboards** area.

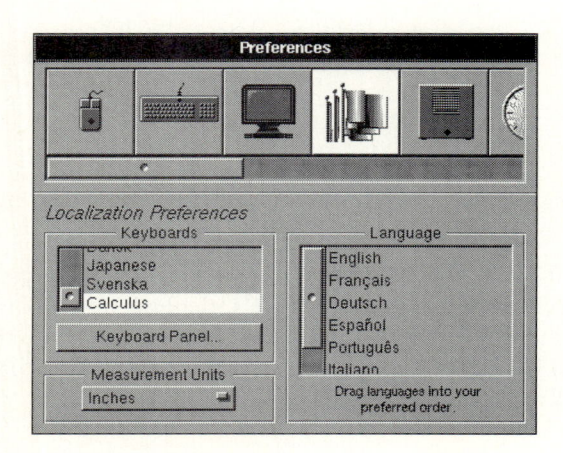

If the installation failed for any reason, start the process again from the top, making sure that all the files you need are copied from the disks.

Provided that there were no errors during the installation, Calculus&*Mathematica* will work with your NeXTStep computer. To make everything run as smoothly as possible, a few preferences need to be changed from within *Mathematica*. The next section deals with these changes and some *Mathematica* basics specific to NeXTStep computers.

Mathematica for NeXTStep basics

Now that the lessons are set up, start *Mathematica*. You can do this by finding the folder called **Mathematica.app**, usually located in **/LocalApps**. Click once on the **Mathematica.app** to make its icon appear in the browser window. You can drag the icon over to the dock at the right so you can find *Mathematica* easily whenever you want. Now double-click on the icon to start *Mathematica*. Later on, you will probably launch *Mathematica* by double-clicking on the icon for the notebook you want to work on.

The icon in the lower left corner can be dragged to the program list on the right side of the screen. Now, whenever you want to run *Mathematica*, just double-click on the icon. Alternatively, you can double-click on a *Mathematica* Notebook to start *Mathematica*, but this is not suggested for the first time as some things have not yet been covered.

Once *Mathematica* is running (it takes a few seconds to load, depending on your system), you will see a document like the following:

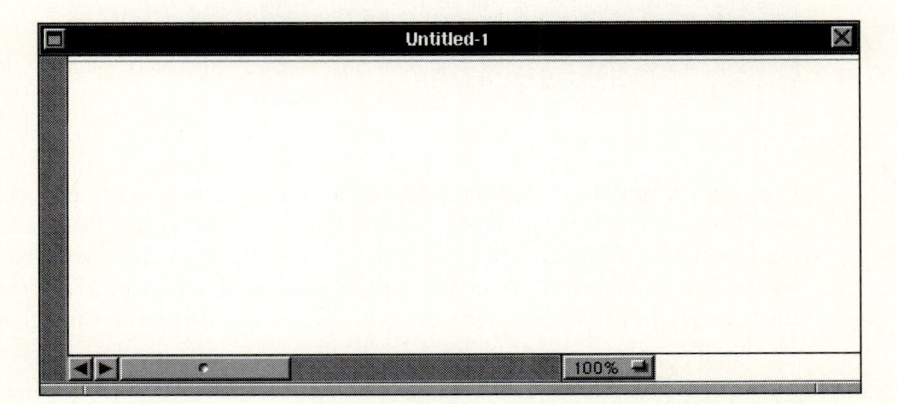

This document, called a Notebook, allows you to enter mathematical expressions and *Mathematica* programs for evaluation, as well as to enter standard text like any good word processor. (Historical note: It is this ability to combine text and expressions that made Calculus&*Mathematica* possible. Without it, you'd be sitting and staring at a blackboard right now.) Notebooks are made of cells, which you cannot see right now, because the whole Notebook is only one cell at the beginning. For more on the basics of *Mathematica* in general, please see the corresponding section of this guide.

What now needs to be done is to change your local settings to make Calculus&*Mathematica* easy to use. This is done by changing the default settings for *Mathematica* to reflect the use of the Calculus font. From the **Info** menu, select **Preferences**, then **File Locations**. The screen will now look like this:

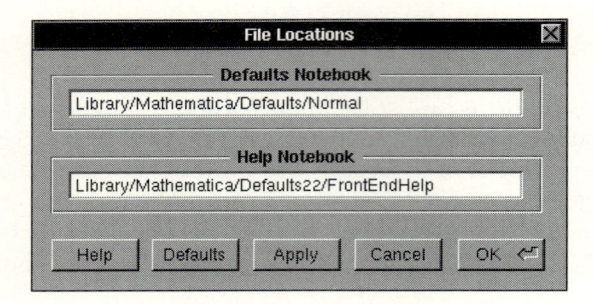

In the field for **Default Notebook**, highlight the word **Normal**, then type **CalcMath**. The window will now look like this:

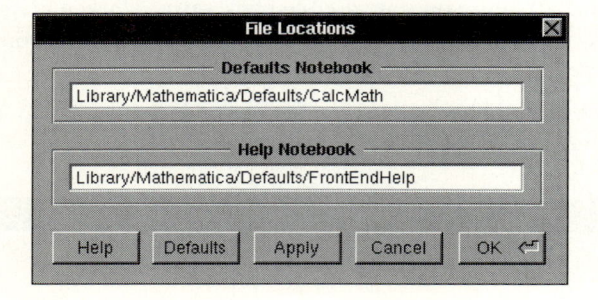

Click on the **Apply** button, then click on **OK**. The preferences are now changed, so any new Notebook you create will automatically be in the style of Calculus&*Mathematica* Notebooks. This allows Notebooks you create to be moved from NeXTStep computers to Microsoft Windows and Macintosh versions of *Mathematica*, with little or no change in appearance. See the section on file transfer for more information on this process.

> Now go have fun. Go to the **0.FeelofMathematica** folder and read the **Greeting**. Then start the **FeelofMathematica** lessons, **Numbers**, **Algebra**, and **Plots**, and work your way through them doing everything they suggest. After that, you'll be comfortable with *Mathematica*, and Calculus&*Mathematica* lessons.

Calculus&*Mathematica* with Microsoft Windows

**by Corey Mutter,
Calculus&*Mathematica* Development Team**

Congratulations! You have selected Calculus&*Mathematica*, the most exciting calculus course on this or any other planet. The goal of the course is, of course, to help you understand mathematics. One of the tools we'll use to aid you in your quest for calculus knowledge is a computer program called *Mathematica*. If your computer is IBM-compatible, you'll need to use Microsoft Windows as well. This guide is intended to help you get comfortable with your computer.

Communicating with a computer can be like communicating with a member of the opposite sex. Much of the time, you and your machine are speaking different languages. In this guide, there'll be a lot of computerese. It's not essential that you memorize them. However, you'll probably pick up some of them by using them. If you know some of the lingo, you'll be able to understand your machine better and communicate with it in a clearer manner. Refer to the glossary when you run into a term that you don't know. You'll be able to understand computer gurus better (when they try to tell you what's wrong), and you and your machine will have a happier relationship. I swear.

This is intended to be a learn-by-doing guide. Ideally, you should be sitting at your computer and playing around with it as you read this. (If you're already familiar with Windows, you can just skim over the Windows stuff.) Don't be afraid to experiment. The machine is just a tool like any other, and it's under your control. Go ahead and turn it on. Windows should start up when you turn it on. (If it doesn't, you'll probably see something like `'c:\ >'` and a blinking line. If you do, type `'win'` and press Enter.) You should see a combination of windows and/or icons on the screen (Figure 1). A window is just a chunk of the screen that is independent of other windows; you can have a window for your word processor, another window

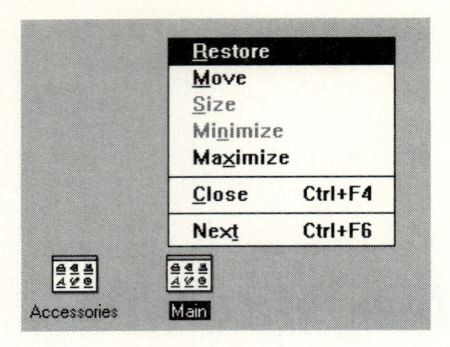

Figure 1. System menu.

for Paintbrush to draw pictures, and a bunch of other windows for a bunch of other stuff, all at the same time. An icon is a little picture that represents something (i.e., a program or a window).

There should also be an arrow. Try moving the mouse around and you'll see that the arrow (the mouse pointer) scoots around the screen as you move the mouse. Go ahead and practice moving the mouse pointer around. If you run out of space on the desk, just pick up the mouse and move it. Moving the mouse while it's not touching the desk doesn't move the pointer on the screen. Now move the pointer so that it's over top of one of the icons. (If there aren't any icons, you can make a window into an icon by moving the mouse pointer over its minimize box (see Figure 2), pressing, and releasing the left mouse button.) Press and release the left mouse button. This is called clicking. Figure 1 illustrates what you should see—the system menu.

A menu (just like in many reputable restaurants) gives you a list of choices. Click on **Move**. ("Click on that" is short for "move the mouse pointer over that, then press and release the left mouse button.") Now press a couple of the arrow keys on the keyboard. The icon will move around the screen. Move it to somewhere that looks aesthetically pleasing to you, then hit **Enter**. There are also other menus; most programs' windows have a menu bar at the top that lets you pop up other menus to do all kinds of stuff.

Move the mouse pointer over top of the icon. Press the left mouse button, but don't let go. Move the mouse around while you're holding the button down. This is called dragging. As you drag the mouse around, the icon will scoot around the screen along with the pointer. You can move a whole window around like this: Just drag around the title bar. Dragging is used a lot in Windows—either to highlight stuff or to move stuff around.

There's another basic mouse operation: double-clicking. To try it out, move the mouse pointer over an icon (if there's no icon, make one; see Figure 2). Now, click twice rapidly (in the space of one second or so). The icon should expand into a window. (Clicking on **Restore** on the system menu will do the same thing.) Double-clicking will highlight (or select) an object and perform some kind of standard action on it (for icons, restoring them into windows). Go ahead and double-click on all the icons you see to make them into windows. If they overlap, don't worry; clicking anywhere inside a window's boundary will move it to the top of the pile and make it the active window. The icons inside the windows don't represent other windows; they're programs. You can start up any of these programs by double-clicking on its icon. More on that later.

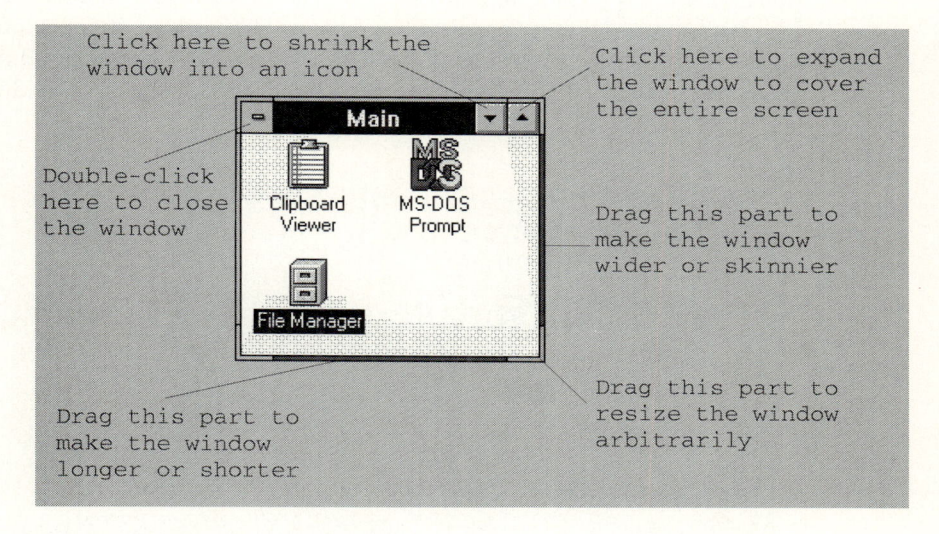

Figure 2. Resizing a window.

There are a few other shortcuts for common window operations. (Refer to Figure 2 to see what these buttons look like.) The downward-pointing arrow at the top right of the window is called the minimize box. Clicking on it will shrink the window into an icon. The upward-pointing arrow to its right is the maximize box; clicking on it will expand the window to cover the entire screen. Clicking on the maximize box of a window that's already been maximized (it'll have two arrows in it) will shrink the window to the size it was before it was maximized. The box on the top left with the line in it is called the system menu box. Clicking on it will bring up that system menu you saw earlier. Double-clicking on it will select **Close** from the system menu, which will ordinarily get rid of the window altogether. (Program Manager's group windows, the ones you're fooling around with now, are an exception; for them, **Close = Minimize**.)

Take a look at one of the windows. Notice how the border around it is divided into sections. There's a good reason for that. If you drag the border around, that part of the border will move while the rest of the window stays put. What you'd be doing is changing the size of the window. Go ahead and try it; use Figure 2 as a guide.

In the process of resizing windows, you may have made a window too small to show all of its stuff at once. If you did, you might've noticed that a bar with arrows appeared (on the right if the window was too short, or on the bottom when the window was too skinny). This is called a scroll bar; what it does is called, surprisingly enough, scrolling. Here's how it works: Suppose you have a window that's too short to show everything that's in it (like a word processor window with a five-page paper in it). The scroll bar on the right will have an arrow pointing down and an arrow pointing up. Clicking on the down arrow would cause the window to scroll down, letting you see what's below the bottom of the window. Likewise, clicking on the up arrow would scroll the window up. Go ahead and try making a window really short (or skinny) and scrolling around in it. Figure 3 is the guide to scrolling.

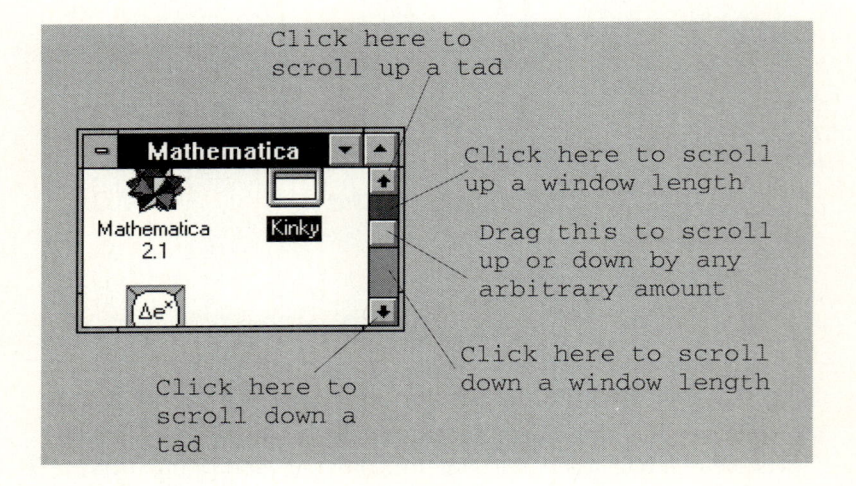

Figure 3. Guide to scrolling.

Now you should know the basic Windows operations. Resize and move the Program Manager windows around until they're arranged in a way that's aesthetically pleasing to you. You may want to minimize some of them as well.

There's an old computer guru saying, "If all else fails, read the instructions." Windows has built-in instructions. To see them, try clicking on **Help** in the menu bar of whatever window you're using. It'll give you a bunch of options for invoking Windows Help, which is pretty friendly. Take a look at it sometime.

The basic task of any computer is to process information. You feed it some stuff, it does something to the stuff, and it gives you back new stuff. For the computer to process information, it has to drag it into its memory banks. One big problem with current computer memory is that it loses its contents when you turn off the power. So, while you're, say, working on a paper with your word processor, it's being stored in your machine's memory. When you turn off the machine, the paper is gone! So, how do you keep it around long enough to print it out?

Disks are the solution to this problem. Once you put data (information) onto a disk, it stays there till you erase it. The chunks of information stored on your disks are called *files*. So the normal operation of a program goes something like this: Load some information into memory (open the file), do whatever needs to be done, and then stick the information back onto the disk for later use (save the file). Then you can work with the file some more later. That's how you'll be doing your lessons in Calculus&*Mathematica*.

Most computer systems use a few different types of disks. You probably have a disk drive for 3.5-inch floppy disks (the little ones in hard plastic cases) or 5.25-inch floppy disks (the soft ones). And, your computer has a hard disk drive. A hard disk is a large-capacity, built-in, fixed disk. On most systems, the hard disk holds Windows itself and any big programs that you keep handy (like *Mathematica*). Floppy disks are good for storing smaller chunks of information (like your lessons). You can carry them around with you.

Disk drives on Windows machines are identified by letters. So, you just give Windows the letter of the drive to use. By tradition, your hard disk is almost always drive C. Drive A is your primary floppy disk drive, and if there's another floppy drive, it's B. Any disk you use has to be set up before you use it. This process is called formatting, and it's akin to putting up bookshelves in a library. The disk needs to be formatted so that the operating system can organize your files and retrieve them efficiently. Your hard disk, of course, will already be formatted; floppy disks, however, usually come from the store unformatted.

Formatting a disk will erase anything that's already on it. Therefore, if you format a disk that contains the only copy of your work, then it's gone forever. To format a disk, double-click on File Manager's icon (it's probably in the **Main** group). You'll see a window with a bunch of file names in it (probably on the hard disk). To look at what's on your floppy disk, shove the disk into the appropriate drive and click on the icon for that drive (Figure 4).

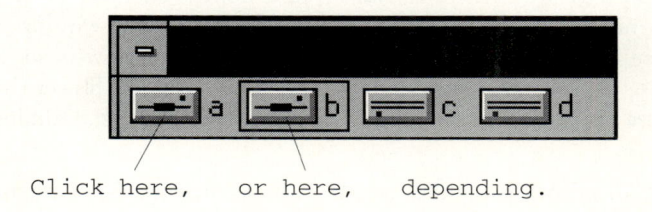

Figure 4. Selecting a drive in File Manager.

If the wrong drive lights up, click **Cancel** when the error message comes up, and try the other drive. If the right drive lights up but there's still an error message, the disk needs to be formatted. (Your machine can't read it. It might be formatted for a Mac or other type of computer.) If files show up, look at them to make sure you don't want to keep them. Once you're satisfied, click on **Disk** on the menu bar. Then click on **Format Disk** on the menu that pops up. You'll get a dialog box (a small window that either gives you information or wants input from you (see Figure 5). Check to see that the drive letter listed after **Disk In:** is the same drive that your disk is in. If it isn't, click on the arrow that's to the right of that box, and then click on the right drive letter. Also, make sure that those two check boxes down at the bottom, **Make System Disk** and **Quick Format**, are both unchecked. If one of them has an **X** in it, click on the box to remove the **X**.

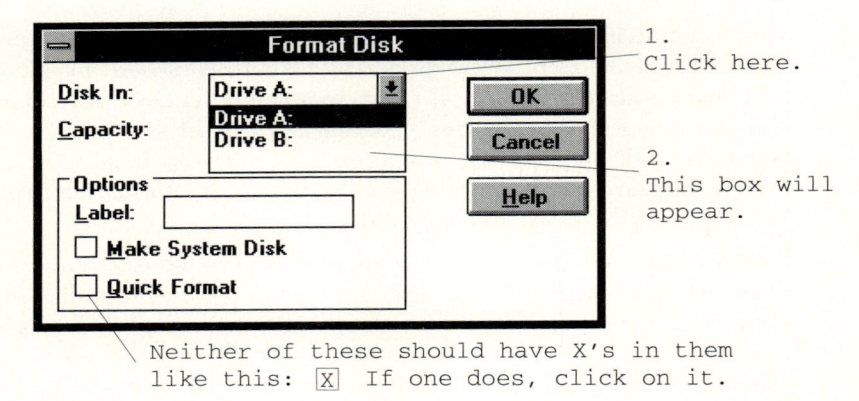

```
1.
Click here.

2.
This box will
appear.

Neither of these should have X's in them
like this:  X   If one does, click on it.
```

Figure 5. Format Disk dialog box.

When you're satisfied, click on **OK**. A warning message will appear, asking if you're sure that you want to format this disk. If you're OK with it, click on **OK**. The disk will now get formatted. When it's done, a dialog box will pop up telling you how much space is on the disk and asking you if you want to format another.

Now you have your own disk that you can use to keep stuff on. You can play with *Mathematica* and save the fruits of your labor. So, go ahead and shrink File Manager down to an icon (click on its minimize box), find *Mathematica*'s icon (probably in a group called **Mathematica**), and double-click on it. A window will pop up with a nice little title screen, and then you'll get a window with a bunch of blank space in it. This is *Mathematica*. All you need now is some information to process. So, click on **File|Open**. (This is short for "click on **File** on the menu bar, and then on **Open** on the menu that pops up.") You should get a dialog box that looks like Figure 6.

On a disk, in addition to files, you can have directories. Directories (sometimes called folders) help you organize your files. You can put all of your lesson files in a directory called '**lessons**', for example. They're also handy for file naming. If you have a file named '**readme.txt**' in a directory called '**scorch**', you can also have a file called '**readme.txt**' in a directory called '**wolf**'. You just can't have two files with the same name in the same directory. A directory can also contain other directories, so on most hard disks, you end up with a directory tree. Every disk

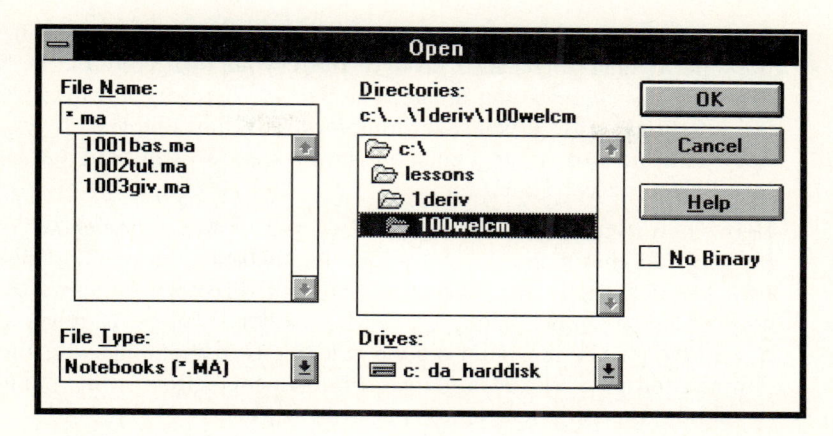

Figure 6. Opening a file.

has a root directory (on the right-hand side, labeled 'c:\'). If you double-click on the root folder, you'll see all of the first-level folders on the disk. On this example system, the lessons are stored in the 'lessons' directory, and the file we want is 'c:\lessons\1deriv\100welcm\1001bas.ma'. (The .ma at the end signifies that the file is a *Mathematica* Notebook. The backslashes are the standard way of writing a path on Windows machines. A path indicates not only the name of a file but also what directories it's part of.)

So, you would double-click on 'c:\', then on 'lessons', then on '1deriv', then on '100welcm' on the right-hand side. If, once you double-click on 'c:\', you don't see the lessons directory, you might need to scroll down. If there are more directories than will fit in that box, you'll be able to scroll through the names with the scroll bar. The left-hand side lists all of the files in the current folder. (If there's a bunch of them, you might need to scroll through them too.) Go ahead and find the folder containing the first lesson. (The path might be different on your system. If you're not sure, ask someone who knows.) Click on '1001bas.ma' (or whatever the first lesson is called), and click on OK. The file will get loaded into *Mathematica*.

When you open up the lesson, a dialog box should pop up asking you if you want to evaluate initialization cells. Say Yes to this. Initialization cells are in the lessons to turn off some of the more annoying error messages that *Mathematica* generates, and to set up some easier ways for you to do things (like draw vectors). Always say Yes, unless you know that you're not going to run any commands. If you say Yes, *Mathematica* will load in its Kernel from disk. (The Kernel is the part of *Mathematica* that does the math. The Front End is the part of *Mathematica* that displays Notebooks.)

> When you tell *Mathematica* to evaluate its initialization cells, all of the closed groups will get opened. Fix it. Click on Edit|Select All Cells, and then on Cell|Close All Subgroups.

You'll probably want to maximize the window so that you can see as much as possible at once. You also need to make your own copy of the file, so that you can save any changes you make. The lessons directory is where fresh copies of the lessons should stay. If you save your work into that, and then if you mess something up, you'll have to try to figure out how to get a fresh copy of the lesson off your installation disks. That's no fun.

Here's how to make your own copy of a lesson: Click on **File|Save As**. You'll see a dialog box pop up that's similar to the one for **File|Open** (Figure 7). The example system here has a folder called '**wkspace**' in the root directory for users to save a working copy of lessons. Click in the box under '**File Name**', delete whatever's there, and type in a name for the file. (It's a good practice to personalize your file names, so that you can tell your files at a glance from someone else's.) The name you type must be eight characters or less, no spaces or symbols.

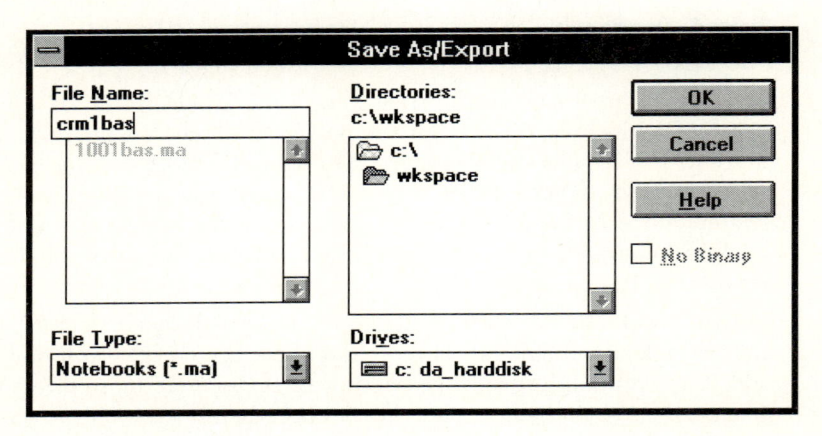

Figure 7. Save As dialog box.

Click on **OK** when you get the file name typed in. You don't have to tack on the **.ma**; *Mathematica* does that for you. (The **.ma** at the end tells the system that the file is a *Mathematica* Notebook.) Now, the name you typed will appear at the top of the *Mathematica* window, and you'll be left there, with the first lesson in front of your eyes. Scroll down until you can see the first problem. Go ahead and start working—just do what the lesson tells you to do. It's pretty self-explanatory.

Now that you've gotten some stuff done, you probably want to save it. (If you don't save it and you exit *Mathematica* or turn off the power, what you did will be irretrievably lost.) It's a good idea to save every so often (oh, a half hour or so is a good stretch of time). Sometimes *Mathematica* will exit itself without giving you a chance to save your work. This is called a crash and can be caused by bugs in *Mathematica*, or, to a lesser extent, bugs in the operating system, problems with the computer's electronics, and power outages.

Click on **File|Save**. The file will get saved for you, without asking for a name. *Mathematica* will just update the file on the disk with the changes you made since the last save. If you want to save the file under a different name (like you did to create the copy of it), use **File|Save As**. If you make changes to your *Mathematica* Notebook and then try to exit *Mathematica*, it'll stop and ask you if you want to save your changes. Clicking on **Yes** will let you save, clicking on **No** will trash the changes that were made to the file, and clicking on **Cancel** lets you take back the **File|Exit** operation. This is a safety feature that keeps you from losing changes that you made. Click on **No** only if you're sure that you don't want what you did saved. Otherwise, hit **Yes** or **Cancel**.

Mathematica includes a set of commands (common to most Windows programs) that will save you an incredible amount of typing. These are **Edit|Cut**, **Edit|Copy**, and **Edit|Paste**. Windows has a clipboard that you can store stuff on. **Paste** takes whatever is on the clipboard and puts it into your file, right where the insertion point is (either the blinking cursor or, in *Mathematica*, the horizontal line), just like you typed it in. **Copy** takes whatever you highlight and copies it to the clipboard. **Cut** takes whatever's highlighted and moves it to the clipboard, removing it from your document. Whenever something's put in the clipboard, it replaces what was on the clipboard before. It can only hold one chunk of data at a time. Even so, it's still mighty useful. Try it out:

Open up a Notebook. Highlight a problem (by clicking on the cell bracket to its right). Click on **Edit|Cut**. The problem you highlighted should disappear. Now click on **File|New**. A blank, untitled Notebook should appear. Now hit **Edit|Paste**. The problem that disappeared from the original Notebook should reappear in your new Notebook. When more than one Notebook is open, you can move around between them using the Window menu. The Window's menu will list each Notebook that's open. Clicking on the Notebook's name will bring its window to the top of the window pile.

The first Calculus&*Mathematica* lesson should have taught you how to type *Mathematica* commands. (In case you missed it, you move the mouse pointer until it becomes a horizontal I-beam, click, then type.) You'll want to type things other than commands, though. Calculus&*Mathematica* places a lot of emphasis on explanations. You'll want to explain what you've done in your homework, and you won't want to make your explanation a *Mathematica* command. (If it got evaluated, that might confuse *Mathematica*. For example, it knows what `Expand[(x-1)^10]` means, but probably is a little fuzzy on what `That looks neat` means.) *Mathematica* has a solution to this. You can have different types of cells. You've seen Title cells (the big Calculus&*Mathematica*) in the beginning, Input cells (where your commands go), and Text cells (words to explain what's going on). Here's how to create some Text cells.

Create a new Notebook (**File|New**). Type something (anything)—a new cell will appear with the stuff you typed in it. Highlight its cell bracket by clicking on it. The cell you created will be **Input**. Look up on the top left corner of the *Mathematica* window, until you see where it says **Input** (Figure 8). Click on the downward-pointing arrow to the right of **Input**. A list of styles will appear. Scroll up the list until you see **Text**. Click on it.

50 **Calculus&*Mathematica* with Microsoft Windows**

Click here first.

Then click here till you see **Text**. *Click on* **Text**.

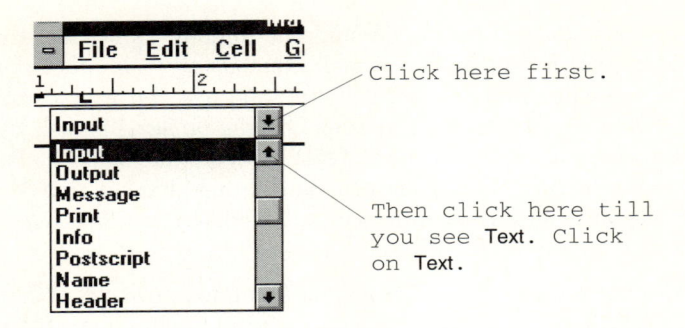

Figure 8. Changing a cell's style.

The stuff you typed will change (it'll probably get taller and turn blue). Now it's text, and can't be run as a *Mathematica* command. (Try it!) Go ahead and type a couple other lines of text. There's another good use for Text cells. You'll need to type mathematical symbols. To learn how to do that, check the section of this book called How to Type Using Calculus Font on Macintosh, NeXT, and Windows Platforms.

Ever wonder what all those buttons at the top of the *Mathematica* window are for? Now's the time to find out. It's called a toolbar, and it's designed to save you some work. Each of those buttons represents a common operation (like **Copy**, **Paste**, animate a graph, run a command, etc.). Clicking on one of those buttons has the same effect as selecting the corresponding menu item. Figure 9 shows what the more heavily used among them do:

One more thing you need to know how to do is manage your files. Windows has a program that lets you do this the easy way—File Manager. Double-click on File Manager's icon to get a look at how it works. (You might need to minimize *Mathematica*, or you can **File|Exit Mathematica**.) You've seen this program before—when you were formatting a floppy disk. The big things you need to know are how to copy, rename, and delete files.

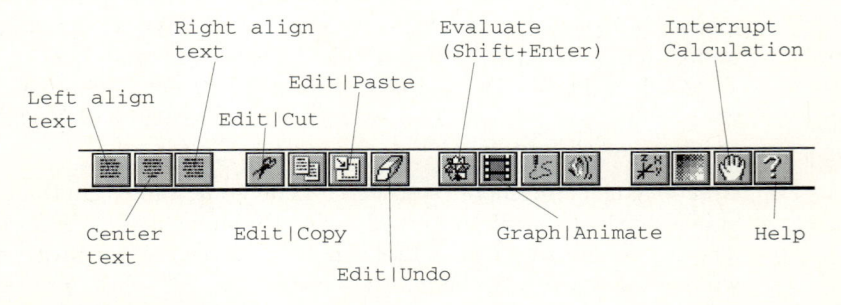

Left align text — Center text — Right align text — Edit|Cut — Edit|Copy — Edit|Paste — Edit|Undo — Evaluate (Shift+Enter) — Graph|Animate — Interrupt Calculation — Help

Figure 9. Toolbar commands.

First, copy the file you saved in the workspace directory onto your floppy disk. To do that, first highlight the workspace directory on the left-hand side. (Do this the same way you found the right directory for opening a *Mathematica* Notebook.) The files in it should appear on the right side of the window. Suppose you saved a file called **1011bas.ma**. Two files will appear—**1011bas.ma** and **1011bas.mb**. Why's that?

The **.ma** file is the actual *Mathematica* Notebook. It contains all of the information needed to reconstruct the Notebook. The **.mb** file is an auxiliary file; it contains information like where the cell brackets are and what the pictures look like. *Mathematica* can calculate this stuff based on what's in the **.ma**, but it's quicker to read it from the **.mb**. However, if the **.mb** becomes invalid or gets renamed in the wrong way, it can screw up your Notebook. Therefore, you'll probably be deleting these files most of the time. So you don't have to save the **.mb**'s on your disk, just the **.ma**'s.

Click on the **.ma** file to highlight it. Now, drag it up into the appropriate drive icon. (Make sure your disk is in the appropriate disk drive.) The file will get copied to your floppy disk. That's all there is to it. Then, if other folks are likely to use your computer, you probably should delete the file in the workspace. Click on the **.ma** file, and then hold down the **Control** key and click on the **.mb** file. Both of them should now be highlighted. Then, click on **File|Delete**. Windows will ask you if you're sure you want to get rid of them. If you say **Yes**, the files are gone forever, and the disk space they used up is freed again.

Sometimes you'll want to change a file's name. To do this, click on the file and hit **File|Rename**. Windows will ask you what you want the new name to be. Type it in, remembering that it can only be eight or fewer letters long. Then, type a period and the same extension as the original (i.e., **.EXE**, **.MA**, **.TXT**, etc.). If you change the extension, how will you (or the system) remember what kind of file it is? You won't. Clicking on **OK** will rename the file. Now you know about all the basic File Manager operations.

Installation

Installing Calculus&*Mathematica* on your hard disk is pretty easy. Just shove Disk 1 into your disk drive. Run the program **INSTALL.EXE**. (You can use either Program Manager's **File|Run** command, or double-click on the file in File Manager.) The lessons will get copied to your hard disk, as well as our font, default styles sheet, typing helper, and anything else you need to use Calculus&*Mathematica*.

Look for a file called **README.TXT**. It will tell you special things you need to know, like how to install the font, which font to install, and where to put special files. One of the things we recommend there is that you have the most recent version of *Mathematica* available. At this printing, that is version 2.2.2.

If you are running a version of *Mathematica* numbered lower than 2.2.2, you will need to use the postscript font provided here. In order to do that, you need **Adobe Type Manager** for Windows. For folks who don't have ATM, we have arranged a special deal. You may order ATM for Windows from Adobe Systems, Inc. by mail to: P.O. Box 6458, Salinas, CA 93912-6458, by phone at 1-800-521-1976, or by fax at 408-655-6073. Total price is $29.00 plus a $7.50 handling charge and tax where applicable. Please mention offer code P-49-12-R.

If you are running a version of *Mathematica* numbered 2.2.2 or higher, you can use the TrueType font provided in the distribution, and install it with the fonts control panel in Windows.

Giving *Mathematica* enough room

One more thing you do need to know to set up *Mathematica* on your computer is how to create virtual memory. *Mathematica*'s Kernel is HUGE. It takes many megabytes of memory just to load it. To actually do math takes even more memory. Most computers don't have all that much, so where will you get that extra space? Virtual memory is the answer.

Virtual memory (also known as a *permanent swap file*) allows you to take a chunk of your hard disk and set it aside to be used as memory. As far as programs are concerned, it's just like regular memory. Start up Control Panel. Double-click on **386 Enhanced** to see some options. The dialog box has a button marked **Virtual Memory**. Click on it.

You'll get another dialog box that tells you what drive your swap file is on, how big it is, and whether it's permanent or not. Permanent is better, because it's faster. If you think it's not big enough, you need to make it bigger. The sum of your real memory and virtual memory probably shouldn't be less than 12–15 MB. If the settings are good, click on **OK** and forget about the rest of this. If they aren't, click on **Change**. You'll get yet another dialog box.

You'll have two lists—one of hard drives and one of swap file types. Select the hard drive you want to have the swap file on, and select a swap file type. Windows will give you three figures for the drive—the free space, the maximum swap file space, and the recommended size. Windows won't make a permanent swap file that's bigger than the recommended size because a permanent swap file has to be all in one continuous chunk of hard drive space. The recommended size that Windows gives you is the biggest single chunk of free space. You can consolidate the free space on your drive with a defragmentation program to get more contiguous space. After you have done everything and set the swap file type and size, make sure the **Use 32-Bit Disk Access** box at the bottom is checked. Click on **OK**. You'll need to restart Windows for the changes to take effect. That's it.

Oh, yes, by the way, if you use Stacker, MS-DOS 6 DoubleSpace, or another disk-doubling utility, you need to be aware of this: A permanent swap file MUST be on the UNcompressed drive. Check the utility's documentation.

> Now go have fun. Go to the **0.FeelofMathematica** folder and read the **Greeting**. Then start the **FeelofMathematica** lessons, **Numbers**, **Algebra**, and **Plots**, and work your way through them doing everything they suggest. After that, you'll be comfortable with *Mathematica*, and Calculus&*Mathematica* lessons.

How to Type Using the Calculus Font on Macintosh, NeXTStep, and Windows Platforms

If you've looked at the printed text of our course, you'll notice several formatted mathematical expressions, like integrals, square roots, and fractions. These are typed in the Calculus font.

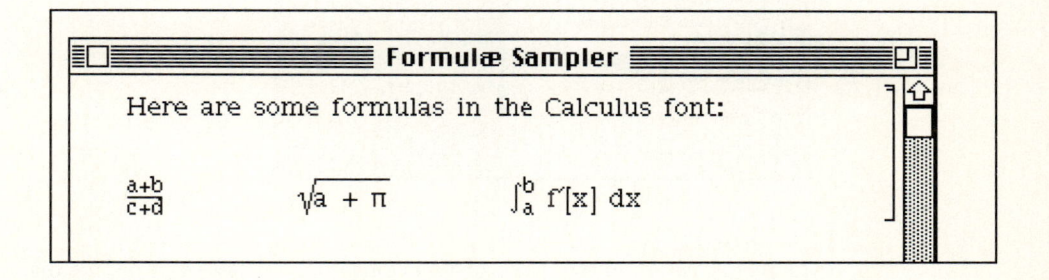

```
┌────────────────── Formulæ Sampler ──────────────────┐
│  Here are some formulas in the Calculus font:       │
│                                                     │
│     a+b              √a + π         ∫ᵇₐ f'[x] dx    │
│     ───                                             │
│     c+d                                             │
└─────────────────────────────────────────────────────┘
```

Mathematica formulas typed in the Calculus font differ from the standard notational practice in that there is no Greek letter except for π. Subscripts and superscripts exist as characters on their own, not on-screen modifications of other characters. This limits superscripts and subscripts (as well as fractions) to one level. Only lowercase letters, digits, ∞, and π can be subscripts or superscripts, with the exception of capital R and capital C, which can be subscripts.

There are two ways to enter formulas using the Calculus font. One way is to enter the text directly, with special keystrokes for superscripts, subscripts, and other special characters. The other way is to use a nifty little *Mathematica* command we wrote that actually takes the code from *Mathematica* input cells and prints them in the Calculus font.

Typing directly into *Mathematica*

The general way of typing in the Calculus font on Mac, or NeXTStep, or in Windows is almost the same on each machine. The only differences are in the actual keystrokes used to get the special characters, the subscripts, and the superscripts.

On a Windows machine, you need to start a program that will transmit the correct keystroke translations. Go to your **Calculus&Mathematica** directory and start the application called **CMType**. Then you can go back to running *Mathematica*, and type the way we describe here.

On a NeXTStep machine, you need to be sure the correct keymapping is active. Get the **Preferences** application by clicking on the clock on the dock, and make it look like the following by clicking on the flag, and then on **Calculus** in the languages choices.

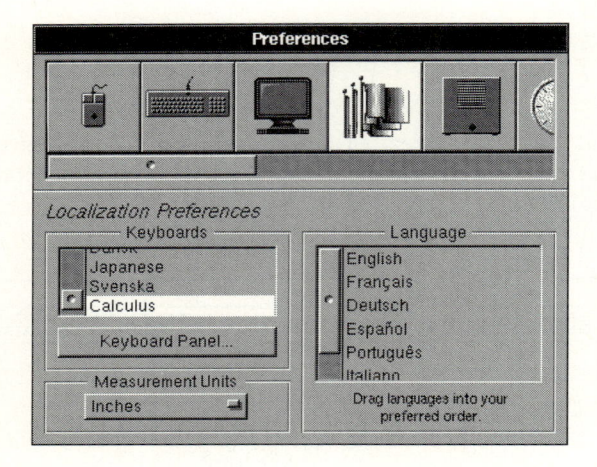

In this section, you will see instructions such as **Option-e**. This means:

- With the **Option** key depressed, type **e**.

Of course, that's an instruction for you if you are running on a Macintosh. On a NeXT machine, the instruction should say **Alt-e**, which means type **e** with the **Alternate** key depressed. On a Windows machine with **CMType** running, the command will be **Control-,e**, which means hold down the **Control** key while typing a comma, and then type an **e**.

Such operations are similar to using the **Shift** key to get capital letters. The **Option** key, **Alt** key, or **Ctl-**, lets you access extra characters that you can't normally see on the keyboard.

Once again, on a Mac, you will also see instructions such as **Option-e**, then **u**. This means:

- With the **Option** key depressed, type **e**.
- Release the **Option** key.
- Type **u**.

(On NeXT and Windows machines, these characters are typed differently. You can see the keystrokes that are needed on the keyboard mappings provided. With **CMType** running on a Windows machine, you can have a keystroke map on the screen as you work. Just find the **CMType** icon that is probably under your *Mathematica* window. Click on it, and request help. Voila!)

Remember, the cell you are typing into must be set to use the Calculus font. We assume that you have installed the font (as well as the keymapping). If you are unsure how to set this, reread the section on Composing a Simple *Mathematica* Document in the introduction to Calculus&*Mathematica* on the Macintosh, or check the *Using Mathematica* manual that came with your copy of *Mathematica*.

Basic typing skills

On a Mac, subscripts are obtained by typing the character you want while pressing the **Option** key. There are exceptions, though:

subscript i	$_i$	**Option-i**, then **i**
subscript n	$_n$	**Option-n**, then **n**
subscript e	$_e$	**Option-e**, then **e**

On a Mac, superscripts are obtained by typing the character you want while pressing the **Option** and **Shift** keys.

On a NeXT, subscripts are obtained by typing the character you want while pressing the **Alt** key. On a NeXT, superscripts are obtained by typing the character you want while pressing the **Alt** and **Shift** keys.

In Windows, subscripts are obtained by typing the character you want after typing **Ctl-comma** (that's **Ctl-,**). In Windows, superscripts are obtained by typing the character you want while holding down the **Shift** key, after typing **Ctl-comma** (that's **Ctl-,**). That's type **Ctl-,** followed by **Shift**-character.

All subscripts (like this: $_{abc}$) are zero width and all superscripts (like this: ^{abc}) are regular width. Zero width means you have to have another character (a space, or perhaps a superscript character) following a zero-width character in order to see the zero-width character.

Extra characters are obtained by the different combinations listed on the keyboard mappings at the end of this section. You really have a lot of flexibility with this font. There are symbols for summation $\sum$, infinity ∞, pi π, and many others.

Here are few examples:

To get $\int_a^b f[x]\,dx$ on a Macintosh,

type $\quad\int\quad$ Shift-3
$\int_a\quad$ Option-a
$\int_a^b\quad$ Option-Shift-b
$\int_a^b f[x]\,dx\quad$ Space, f[x] dx

To get $\int_a^b f[x]\,dx$ on a NeXTStep machine,

type $\quad\int\quad$ Shift-3
$\int_a\quad$ Alt-a
$\int_a^b\quad$ Alt-Shift-b
$\int_a^b f[x]\,dx\quad$ Space, f[x] dx

To get $\int_a^b f[x]\,dx$ on a Windows machine,

type $\quad\int\quad$ Shift-3
$\int_a\quad$ Ctl-, -a
$\int_a^b\quad$ Ctl-, -Shift-b
$\int_a^b f[x]\,dx\quad$ Space, f[x] dx

It looks pretty good. Notice how we did the zero-width character (subscript a) first, then wrote above it with a regular-width character (superscript b) to get the range of the integral. Now let's try something a little more difficult. Here are just the keystrokes for the Mac. You can easily make the changes in keystrokes for the other machines.

To get $\int_{tx+u}^{mx+k} f[x]\,dx$,

type $\quad\int\quad$ Shift-3
$\int_t\quad$ Option-t
$\int_t^m\quad$ Shift-Option-m
$\int_{tx}^m\quad$ Option-x
$\int_{tx}^{mx}\quad$ Shift-Option-x
$\int_{tx+}^{mx}\quad$ Option +
$\int_{tx+}^{mx+}\quad$ Shift-Option +
$\int_{tx+u}^{mx+}\quad$ Option-h
$\int_{tx+u}^{mx+k}\quad$ Shift-Option-k
$\int_{tx+u}^{mx+k} f[x]\,dx\quad$ Space, then f[x] dx

That was a lot of typing for a simple formula, but it looks great!

Algebraic expressions

You can type integrals, fractions, and square roots, and you certainly can type
$$x^3 + a\,x^2 + b\,x + c, \quad e^{-k}, \quad \text{Cos}[x^2], \quad \text{and} \quad (x^8 - 1)/(x^2 - 1).$$

For combinations of subscripts and superscripts, the same rules apply:

$$p[x] = a_0 + a_1\,x + a_2\,x^2 + a_2\,x^3 + \cdots + a_n\,x^n.$$

Shortcut: The `CMForm` command

For most simple expressions, like the first integral above, the above method of typing works fine. In fact, it's what we did to type the whole course for you. On the other hand, you probably think that it takes too long to create these formulas. Well, we thought the same thing. We understood that:

1. Typing formulas is systematic.
2. This can be described in an algorithm.
3. *Mathematica* can do algorithms.
4. Make *Mathematica* do it.

The result is the `CMForm` command. It lets you generate formulas, both simple and complex, using the formatting ability of the Calculus font and the power of *Mathematica*.

To use `CMForm`, you first must load the definition into *Mathematica*. It came with Calculus&*Mathematica*. Here it is on a Macintosh screen (**CMForm.ma** came in the **0.FeelofMathematica** folder. We dragged it out so we could use it easily.):

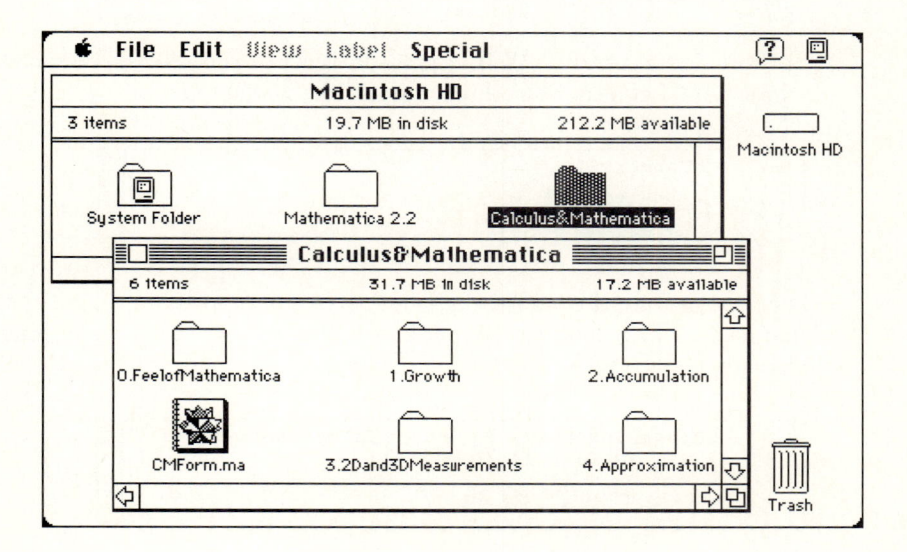

Double-click on **CMForm.ma** to launch *Mathematica* and load **CMForm.ma**. If you are asked, "**Evaluate the initialization cells?**," click on the **Yes** button. These cells set up the definition of **CMForm**. If *Mathematica* does not ask you about evaluating initialization cells, you'll have to manually execute the initialization cells.

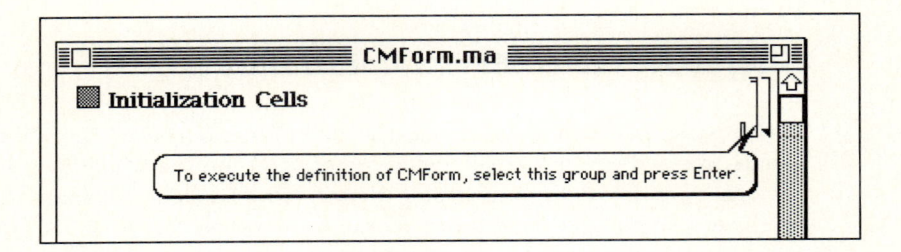

```
┌─────────────────────── CMForm.ma ───────────────────────┐
│ ■ Initialization Cells                                  │
│                                                         │
│     ┌─────────────────────────────────────────────┐     │
│     │ To execute the definition of CMForm, select │     │
│     │ this group and press Enter.                 │     │
│     └─────────────────────────────────────────────┘     │
└─────────────────────────────────────────────────────────┘
```

Before we start using the **CMForm** command, we have to set the Print cell to the Calculus font. This is because **CMForm** uses the *Mathematica* **Print** command. If you are unsure about how to do this, check the *Using Mathematica* manual.

Let's ask *Mathematica* about the **CMForm** command.

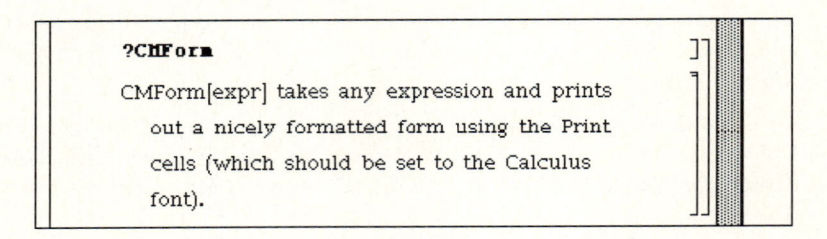

```
    ?CMForm
    CMForm[expr] takes any expression and prints
        out a nicely formatted form using the Print
        cells (which should be set to the Calculus
        font).
```

So how do we use it? Well, all we do is wrap the **CMForm** command around an expression. Let's try something simple, like x-squared.

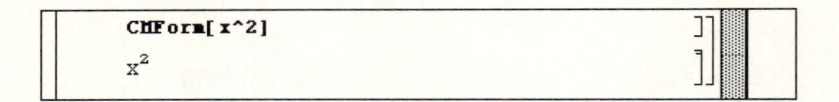

```
    CMForm[x^2]
    x²
```

Now you can use the mouse to select the output expression, copy it to the clipboard, and paste it into another cell, perhaps a Text cell. Let's try something a little more challenging.

```
    CMForm[Sqrt[u] + Pi u^2 + Integrate[u/5,
        {u,(a - 1),(b - 2)}]]
    √u + π u² + ∫ₐ₋₁^(b-2) u/5 du
```

That looks great! And now we don't have to remember the keystrokes for square roots, superscripts, subscripts, integrals, or special symbols like π. How about that?

How to Type Using the Calculus Font on Macintosh, NeXTStep, and Windows Platforms 59

Calculus font: Macintosh keyboard mapping

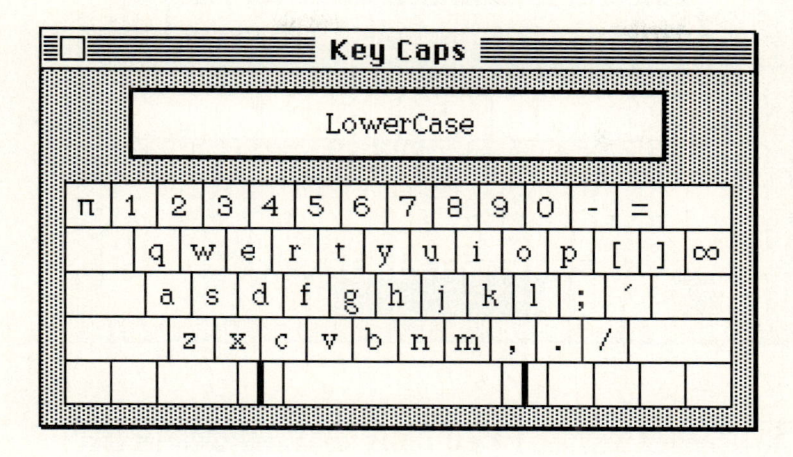

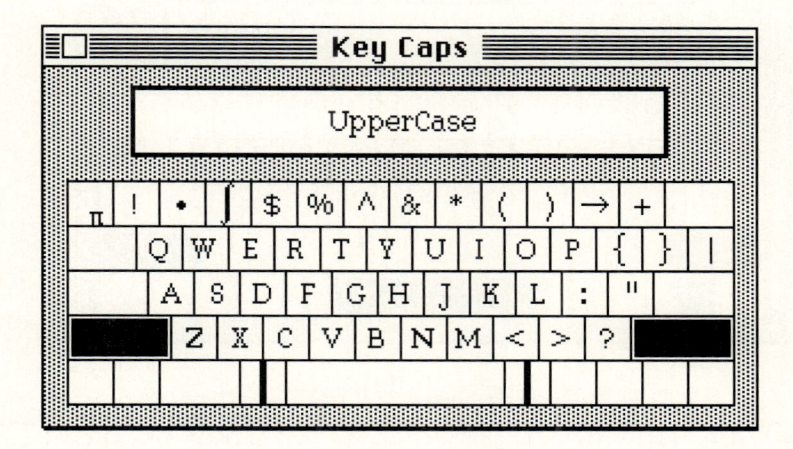

```
╔═══════════════ Key Caps ═══════════════╗
║                                        ║
║        ┌──────────────────────┐        ║
║        │      LowerCase       │        ║
║        └──────────────────────┘        ║
║  π  1  2  3  4  5  6  7  8  9  0  -  = ║
║    q  w  e  r  t  y  u  i  o  p  [  ] ∞║
║     a  s  d  f  g  h  j  k  l  ;  ´   ║
║       z  x  c  v  b  n  m  ,  .  /    ║
║                                        ║
╚════════════════════════════════════════╝
```

```
╔═══════════════ Key Caps ═══════════════╗
║                                        ║
║        ┌──────────────────────┐        ║
║        │      UpperCase       │        ║
║        └──────────────────────┘        ║
║  π  !  •  ∫  $  %  ^  &  *  (  )  →  +║
║    Q  W  E  R  T  Y  U  I  O  P  {  } |║
║     A  S  D  F  G  H  J  K  L  :  "   ║
║  ▓▓▓  Z  X  C  V  B  N  M  <  >  ?  ▓▓║
║                                        ║
╚════════════════════════════════════════╝
```

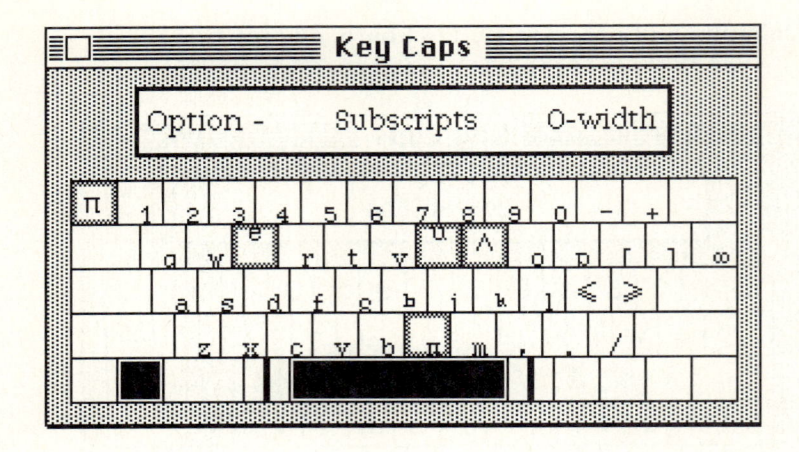

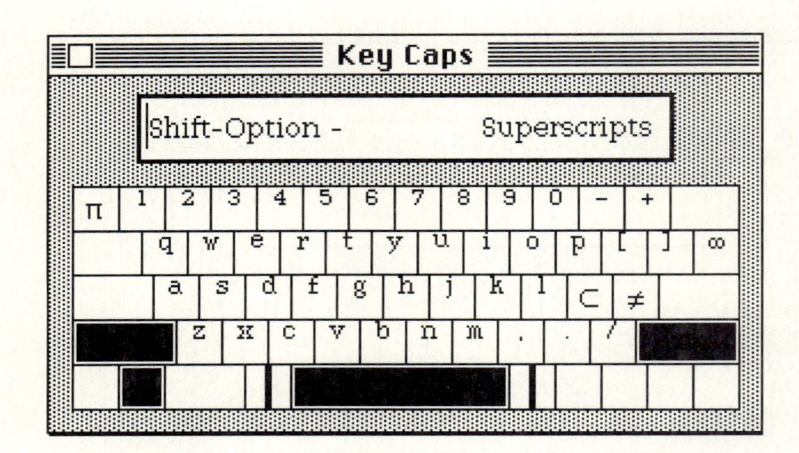

60   How to Type Using the Calculus Font on Macintosh, NeXTStep, and Windows Platforms

	a	e	i	o	u	n	A	E	I	O	U	N	y	Y
O-`	∑	´	^	`	¨		→	/	∩			˛		
O-e	⊂	e	−	]	[		×							
O-i	R	∮	i	»	«		≈							
O-n	∂			~		n	Δ			ǀ		∇		
O-u	∧	=	1	√	u		→	=		²	∪		−	π

Calculus font: NeXT keyboard mapping

62   How to Type Using the Calculus Font on Macintosh, NeXTStep, and Windows Platforms

How to Type Using the Calculus Font on Macintosh, NeXTStep, and Windows Platforms

Calculus font: Windows keyboard mapping

Your keyboard

Shifted

Control+comma, then unshifted

Control+comma, then shifted

Control+period, then unshifted

Control+period, then shifted

Moving Files Among Different Kinds of Machines

As of this printing, Calculus&*Mathematica* courseware is supported on the Notebook Front End versions of *Mathematica* for Microsoft Windows, Apple Macintosh, and NeXT computers. The course Notebooks themselves are platform independent—that is, they can be executed on all versions of *Mathematica* listed above. Moving Calculus&*Mathematica* Notebooks among these platforms is relatively simple (compared to the rest of the computer world!), and this section deals with how to go about the transfer and provides some caveats regarding the process.

Basically, a *Mathematica* Notebook is a simple text document that can be read by any type of computer and interpreted by all versions of *Mathematica* listed above. However, most Notebooks also use binary data, which are highly machine specific and cannot be moved between the Macintosh, PC, and NeXT. These data enable *Mathematica* to open a Notebook more quickly than if it were not present, but the binary data are not essential to the Notebooks in Calculus&*Mathematica*. On the Macintosh, these data are stored in the binary fork of the Notebook, unseen by the user. Following the suggested guidelines for transferring Macintosh-generated Notebooks will assure that there will be no problem with binary incompatibilities. On the PC and NeXT platforms, the binary data are stored in a file with an **.mb** extension to the file name. For instance, a file named **mynotbok.ma** will usually have an associated **mynotbok.mb** file, which is an auxiliary file containing information about rendered graphics cells and other specific information. When transferring Notebooks, be sure not to convert the **.mb** files, because they will not be readable on the new machine.

One of the easiest ways of moving Notebooks among platforms is in a networked computing environment, where different platforms are linked by some form of network hardware, such as modems, ethernet cards, or Apple Computer's LocalTalk cabling. In such an environment, transferring a Notebook to a different computer requires linking to the network and placing the Notebook on the remote computer, since the new machine usually treats the remote computer like it was just another disk drive. Consult the network administrator for specific instructions on how to do this, as procedures vary from one network to another.

Transferring Notebooks without a network usually requires another step, but is not that much more cumbersome than with the network. In this case, the Notebook must be stored on a floppy disk that may be read by both types of computers. Transference between the NeXT and PC computers is easily performed in this manner, since NeXTs are able to read and write DOS-formatted floppy disks. To make the transfer, use a DOS-formatted disk (high-density disks are recommended) and copy the **.ma** version of the file from the original machine (NeXT or PC) to the floppy. Then place the floppy in the target computer and copy the file from the floppy to the computer's hard drive. Transferring files between a PC and a Macintosh follows the same procedure, but the Macintosh requires a separate program to enable the reading of DOS-formatted disks. Two such programs are **PC Exchange** and **Apple File Exchange** (**Apple File Exchange** is included with the system software on many Macintosh systems). Simply follow the instructions on these programs to use a DOS disk in a Macintosh. It might be necessary to change some settings when using these programs, as Macintosh, DOS, and UNIX file systems handle return characters in different ways. If the default settings do not work, try changing the CR-LF (carriage return-line feed) settings. It is also possible to use a DOS disk to transfer Notebooks between a Macintosh and a NeXT, again through the use of extra software on the Macintosh. Additionally, most NeXTs are capable of reading and writing to Macintosh-formatted high-density disks, in which case the procedure is identical to that of sharing between a PC and a NeXT.

One platform difference worth mentioning is the file-naming conventions on each computer. The PC requires that names be in 8.3 format, which means that the file name can contain at most eight characters, a period, then at most three more characters. The Macintosh and NeXT make no such requirement. Be careful when moving Notebooks to and from a PC, as illegal names will be truncated and may overwrite existing files. Additionally, the PC and NeXT often use any characters after the period in a name to link the file to a specific application. Files ending in **.ma** will usually call *Mathematica* on the PC and NeXT computers, but not on the Macintosh. Because of this, after moving a Notebook to a Macintosh, the Notebook must be opened from within *Mathematica* and then saved to ensure that it is correctly linked.

One final note: Although *Mathematica* Notebooks are generally platform independent, a Notebook generated on one platform will not necessarily look identical on another platform. Some basic retouching is usually required in order to make a Notebook look its best after a file transfer.

Trashed Notebooks

Every once in a while, you open a Notebook you've been working on, and it looks brain dead, like this:

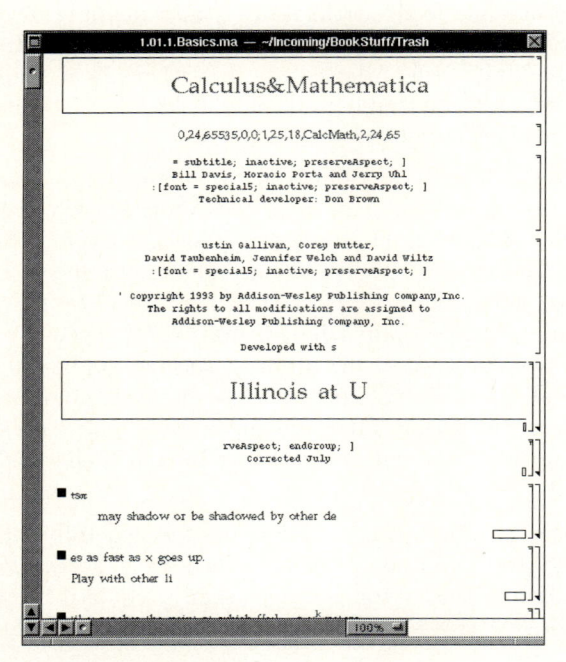

Stop! Don't save

There's a chance that if you catch it soon enough, it will be OK.

Trashed Notebooks 67

Close the Notebook and refuse to save the changes. The problem is probably in that pesky binary fork: the **.mb** part of the file on NeXT or Windows or the hidden binary fork in a Macintosh file. Here's how to try to fix it. Basically, you are going to tell *Mathematica* to ignore the binary bit. On a Mac, pull the file menu down and select **Open Special**.

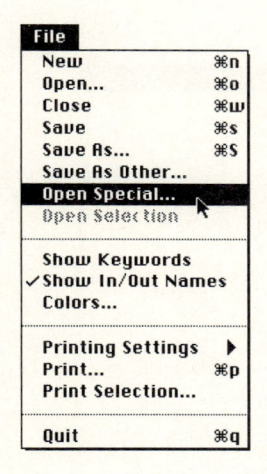

```
File
New              ⌘n
Open...          ⌘o
Close            ⌘w
Save             ⌘s
Save As...       ⌘S
Save As Other...
Open Special...
Open Selection

Show Keywords
✓Show In/Out Names
Colors...

Printing Settings  ▶
Print...         ⌘p
Print Selection...

Quit             ⌘q
```

You'll get a dialog box like this one:

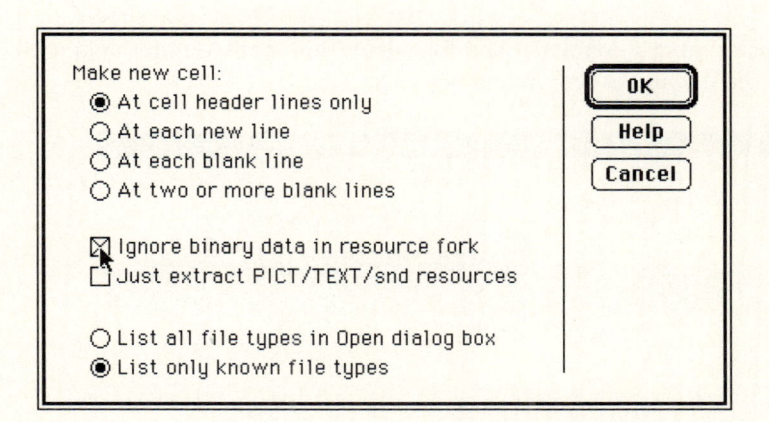

```
Make new cell:
  ● At cell header lines only          [   OK   ]
  ○ At each new line                   [  Help  ]
  ○ At each blank line                 [ Cancel ]
  ○ At two or more blank lines

  ☒ Ignore binary data in resource fork
  ☐ Just extract PICT/TEXT/snd resources

  ○ List all file types in Open dialog box
  ● List only known file types
```

Click on **Ignore binary data**, and say **OK**. Then you'll get a normal open menu from which you should select your file, click **Open**, and hope it comes out looking normal again. In Windows or NeXTStep, the process is similar, but ruffians do one extra step first: Throw the **.mb** part of the file away first, and then go through the same procedure of going to **Open Special**.

68 **Trashed Notebooks**

If it worked, the Notebook should look right again. If it didn't, you'll probably have to go back and retype the stuff. If you are made of really stern stuff, you can try to copy the good bits of the Notebook into a new *Mathematica* Notebook by cutting and pasting. This takes a lot of patience: *Mathematica* Notebooks are ASCII documents. You can open your Notebook with a program that reads and lets you modify text documents, like a word processor. Here's what that mess looks like at the top:

```
                  1.01.1.Basics.ma — ~/Incoming/BookStuff/Trash
(*^
::[     frontEndVersion = "NeXT Mathematica Notebook Front
End Version 2.1";
        next21StandardFontEncoding;
        paletteColors = 128; currentKernel;
        fontset = title, inactive, nohscroll,
noKeepOnOnePage, preserveAspect, cellOutline,
groupLikeTitle, center, M27, N18, O486, R65535, e8, 24,
"CalcMath"; ;
        fontset = subtitle, inactive, nohscroll,
noKeepOnOnePage, preserveAspect, groupLikeTitle, center,
M7, O486, R11051, G11051, B11051, e6, 12, "CalcMath"; ;
        fontset = subsubtitle, inactive, nohscroll,
noKeepOnOnePage, preserveAspect, cellOutline,
groupLikeTitle, center, M27, O486, R32768, e6, 24,
"CalcMath"; ;
        fontset = section, inactive, noPageBreakBelow,
nohscroll, noKeepOnOnePage, preserveAspect,
groupLikeSection, grayBox, M27, O486, R11051, G11051,
B11051, a20, 16, "CalcMath"; ;
        fontset = subsection, inactive, noPageBreakBelow,
nohscroll, noKeepOnOnePage, preserveAspect,
groupLikeSection, blackBox, M27, O486, R11051, G11051,
B11051, a15, 12, "CalcMath"; ;
        fontset = subsubsection, inactive,
```

That's all cell definition and font definition stuff you are looking at. Farther down in the document, you might see this:

```
                  1.01.1.Basics.ma — ~/Incoming/BookStuff/Trash
Steady (negative) growth as x advances from left to right.
Play with other choices of a and b until you get the feel
of a line function.
:[font = subsubsection; inactive; preserveAspect; ]
B.1.a.i) Constant growth rate
:[font = text; inactive; preserveAspect; ]
The most important feature of a line function
        f[x] = a x + b
is revealed by the following calculation.
:[font = input; preserveAspect; ]
Clear[f,a,b,x,h]
f[x_] = a x + b;

Expand[f[x + h] - f[x]]
:[font = text; inactive; Cclosed; preserveAspect;
startGroup; ]
What feature of line functions is revealed by this
calculation?
:[font = special1; inactive; preserveAspect; ]
Answer:
:[font = smalltext; inactive; preserveAspect; endGroup; ]
The calculation reveals that when you take a line function
        f[x] = a x + b,
then you find that
        f[x + h] - f[x] = a h.
This tells you that when x advances by h units, then f[x]
```

That highlighted stuff is stuff you typed in. You can copy and paste it into an input cell in a new Notebook. This process is tedious at best. Most folks have found that rebuilding a Notebook from a fresh copy is quicker and more illuminating.

How could it have happened?

There are lots of ways, of course. The most common is that people who move files around from one place to another get used to just moving the **.ma** parts of the file. If there was already an **.mb** file there with the same name, this tragedy is guaranteed. That can't happen on the Mac, of course, so something different must have happened there. Maybe a couple of bits of information got lost in a file transfer, or maybe the Notebook got opened in a different application, changed ever so slightly, and saved. Maybe you took your floppy disk into the shower with you.

The part you should understand is that the binary part of a *Mathematica* Notebook contains information about where cell brackets should go, things like italics and boldface for text, and any foreign objects that may live in your file. For example, if you insert a picture made by any kind of Paint or Draw program into a *Mathematica* document, it will have to live in the binary fork of the file, since its internal format is machine specific. That's why you don't see photos of lunar landers and computer screens in the electronic Notebooks in Calculus&*Mathematica*.

Another cosmetic problem

Sometime, you will open a Notebook expecting to see something that looks like $\int_a^b e^{tx}$, or $x^2 + 5x + 7$, and see, instead,

$$\downarrow\|\lceil\Leftrightarrow\wr\nabla\S\Downarrow\nabla\S\Downarrow\swarrow$$

Well, there might be several things that cause this. The first one to consider is that the cell isn't a Text cell, or one of the others that use the Calculus font. Change the cell type to Text, or whatever, and see if it looks good. You can do that quickly by selecting the cell bracket and hitting **Command-7** (**Control-7** in Windows) to get it to be a Text cell. If that doesn't work, select that nasty-looking stuff by clicking and dragging, and change its font to Calculus by using the appropriate font change menu stuff for your computer. (It's under different menu choices on each of the three platforms.) If that doesn't do the trick, you have reason to believe that the Calculus font isn't installed or isn't installed correctly. If that's the case, you'll have to go back to the installation stuff and do at least part of it again.

Most of the time, it's the cell problem, not the font problem.

There is another way this can happen when you type into a new Notebook, even in a Text cell: If *Mathematica* doesn't know where to look for the **CalcMath.ma** default files, *Mathematica* won't know about the Calculus font. Check to see, on Mac and NeXT, that in *Mathematica*, the **File Locations** are set correctly as described in the installation procedure.

These represent the most common problems. The next most common typing problem comes from not selecting the appropriate keymapping in the **Preferences** application on the NeXT machine.

Glossary of Computerese

bug:	n.	Design mistake that causes problems.
byte:	n.	A measure of memory or disk storage space.
cell:	n.	The area inside a cell bracket in *Mathematica*. Use of cells allows you to create different styles for different parts of the *Mathematica* Notebook.
character:	n.	A letter, digit, or symbol, or a binary number between 0 and 255.
click:	v.	Press and release the mouse button (usually the left one on PCs and NeXTs).
copy (Edit—):	v.	Make a copy of the highlighted stuff and place it in a temporary storage space, called the clipboard.
crash:	v.	When the computer unexpectedly shuts down, usually due to a problem. Doesn't give you a chance to save your work.
cursor:	n.	The place where what you type is inserted. Usually a blinking vertical line.
cut (Edit—):	v.	Remove the highlighted stuff and place it in the clipboard.
data:	n.	Information.
delete:	v.	Erase.
dialog box:	n.	A small window that gives you information or asks you a question.
directory:	n.	A "folder" that contains a bunch of files. Used to organize files. Separate directories are totally independent, just like file folders in a file cabinet.

72   Glossary of Computerese

disk drive:	n. Part of the computer that reads/writes disks. In Windows, drives are identified by letters; on NeXTs, they show up as different directories; and on Macs, drives have names just like files.
double-click:	v. Click twice in rapid succession.
drag:	v. Move the mouse pointer while holding down the mouse button.
evaluate:	v. Send a *Mathematica* command to the Kernel. The Kernel gives output (calculation result, graph, function definition, etc.).
execute:	v. (a program) Start it up. (a *Mathematica* command) Evaluate a bit of code.
file:	n. Chunk of information on a disk.
file naming:	v. For Windows machines, file names can be eight or fewer characters. Spaces and symbols aren't a good idea. Can be followed by a period and up to three characters. This extension indicates the file's type (i.e., text, *Mathematica* Notebook, Word document, Paintbrush picture, etc.). Macs and NeXTs allow longer file names, with as many dots as you want; however, the last three characters should be .ma for *Mathematica* notebooks.
floppy disk:	n. Small-capacity, removable disk. Comes in 3.5- and 5.25-inch sizes.
folder:	n. Synonym for directory.
font:	n. Style of text.
format:	v. To set up a floppy disk to store stuff on it. ERASES EVERYTHING THAT'S ALREADY THERE, if anything.
Front End:	n. Part of *Mathematica* that shows you the Notebook.
hard disk:	n. Large-capacity, fixed disk built in to your computer. Traditionally drive C on Windows machines, / on NeXTs, and the main directory on Macs.
highlight:	v. Select.
Kernel:	n. Part of *Mathematica* that does math.
kilobyte:	n. 1,024 bytes.
maximize:	v. To expand a window to cover the entire screen.
megabyte:	n. 1,024 kilobytes = 1,048,576 bytes.
menu:	n. List of choices.
minimize:	v. To shrink a window into an icon.
Notebook:	n. *Mathematica* file.
open:	v. To read a file into the computer's memory.
operating system:	n. Underlying control program that keeps track of files on disks, manages your computer's memory, and does other technical things.

paste (Edit—):	v. Take the stuff on the clipboard and insert it where the cursor is in the file you're working on. Leaves a copy of the stuff on the clipboard.
path:	n. A file's name, along with a description of where to find it (what directories it's part of).
program:	n. A set of instructions for the computer.
root directory:	n. The top-level folder on a drive.
run:	v. Same as execute.
save:	v. Update the file on disk with the changes you've made since the last save. Erases the file (if any) that has the name you're saving under and replaces it with the updated version.
scroll:	v. Scoot stuff in a window around so you can see more of it.
select:	v. Drag the mouse over something. Many operations use whatever's selected (like Edit—Cut, Copy, and Paste).

Why Was Calculus&*Mathematica* Written? What Principles Underlie the Course?

The last decade has seen increasingly strong dissatisfaction with the traditional American calculus course. At two national conferences held in the late 1980s, national leaders in science went on record. Here is some of what they said:

Calculus as currently taught is, alas, full of inert material.... The real crisis is that at present [calculus] is badly taught; the syllabus has remained stationary, and modern points of view, especially those having to do with the roles of applications and computing, are poorly represented.
...Peter Lax, Past President of the American Mathematical Society

If a field ever needed to be brought out of mystery to reality, it is calculus.... Calculus is really exciting stuff, yet [the traditional course is] not presenting it as an exciting subject.... Calculus must become a pump instead of a filter in the pipeline.
...Robert White, Past President of the National Academy of Engineering

We no longer ask students to understand. Now it is [algebraic] manipulation pure and simple.... What we're teaching is not only the wrong thing—in that it is not what students will use [outside the calculus classroom]—it is obsolete [because of computers and calculators]. It is like spending all your time in elementary school adding and subtracting and never being told what addition and subtraction are for.
...Ronald Douglas, Dean, State University of New York at Stony Brook

[The traditional calculus course deals with] precisely the ability to do [by hand] the kinds of things that calculators and computers are now doing.... There is a great deal of concern about making sure that freshman courses ... make a significant contribution to broaden the aims of undergraduate education, that they help students learn to think clearly, to communicate, ... etc. There is very, very little in the calculus course of today that does any of these things.
...Lynn Arthur Steen, Past President of the Mathematical Association of America

Lax also gave his assessment of the traditional calculus course:

There is too much preoccupation with what might be called the magic in calculus. For instance, too much time is spent in pulling exact integrals out of a hat, and, what is worse, in drilling students how to perform this parlor trick. Summing infinite series is another topic that has the aura of a magic trick and is overemphasized at the expense of the concept of approximation.... I feel that rigor at this level is misplaced; it appears as an arid game to those who understand it, and mumbo jumbo to those who don't.... Many students have difficulty in grasping the idea that the integral of a function over an interval is a number. The reason is that this number is difficult to produce by traditional methods—i.e., by antidifferentiation—and so the central idea is lost. Numerical methods have the great virtue that they apply universally. When special methods are introduced to deal with each one of the pitifully small class of [differential] equations that can be handled analytically, students are apt to lose sight of the general idea that every differential equation has a solution and that this solution is uniquely determined by initial data.... That today we can use computers to explore the solutions of [differential] equations is truly revolutionary; we are only beginning to glimpse the consequences.

Steen went on to lay out his vision of mathematics courses of the future:

Mathematics-speaking machines are about to sweep the campuses.... The ready availability of powerful computers will enable students to set new ground rules for college mathematics.... Teachers will be forced to change their approach and their assignments. They will no longer be able to teach as they were taught in the pencil and paper era.

Undergraduate mathematics will become more like real mathematics.... By using machines to expedite calculations, students can experience mathematics as it really is—as a tentative, exploratory discipline in which risks and failures yield clues to success. Computers change our perceptions of what is possible and what is valuable.

Weakness in algebra skills will no longer prevent students from pursuing studies that require college mathematics.

Mathematics learning will become more active and hence more effective. By carrying most of the calculational burden of mathematics homework, computers will enable students to explore a wider variety of examples, to study graphs of a quantity and variety unavailable with pencil-and-paper methods, to witness the dynamic nature of mathematical processes, and to encourage realistic applications using typical, not oversimplified, data.

Students will be able to explore mathematics on their own without constant advice from their instructors. Although computers will not compel students to think for themselves, these machines can provide an environment in which student-generated mathematical ideas can survive.

Study of mathematics will build long-lasting knowledge, not just short-lived strategies for calculation. Most students take only one or two terms of college mathematics, and quickly forget what little they learned of memorized methods for calculation. Innovative instruction using a new symbiosis of machine calculation and human thinking can shift the balance of mathematical learning to understanding, insight, and mathematical intuition.

Written in response to the issues raised by Lax, White, Douglas, and Steen, Calculus&*Mathematica* attempts to transform Steen's vision into reality.

Although Calculus&*Mathematica* was written in response to needs of American education today, many of the principles underlying Calculus&*Mathematica* have already been announced by the scientists who made a difference.

It is unworthy of excellent persons to lose hours like slaves in the labor of calculation.
...Gottfried Wilhelm von Leibniz

Some defects ordinarily found in the [educational] method used by mathematicians:

Defect 1: To pay more attention to proof than to evidence, and to try to convince rather than enlighten the mind.

Defect 2: To prove things that do not require proof.
...Antoine Arnaud and Pierre Nicole

[The normal student attempts to fix formal terminology] in his memory because it means nothing to his intelligence.
...Blaise Pascal

Never undertake to prove things that are so evident in themselves that one has nothing clearer by which to prove them.
...Blaise Pascal

[Some facts] can be seen more clearly by example than by a proof.
...Leonhard Euler

[A good calculating machine could] weave algebraic patterns just as the ... loom weaves flowers and leaves.
...Augusta Ada Byron

We must agree that the idea [of formal definition of limit] is not sufficiently clear to use as a basis for a science in which certainty must be founded on evidence, especially when presented to beginners.
...J. L. Lagrange

Is it life, I ask, is it even prudence,
To bore thyself and bore the students?
...Johann Wolfgang von Goethe

All theory, dear friend, is grey, but the golden tree of actual
life springs ever green.
...Johann Wolfgang von Goethe

When a student commences seriously to study mathematics, he believes he knows what a fraction is, what continuity is, and what the area of a curved surface is; he considers evident, for example that a continuous function cannot change its sign without vanishing. If ... [the teacher] says to him: No that is not evident; I must prove it to you, ... what would this unfortunate student think? He will think that the science of mathematics is only an arbitrary accumulation of useless subleties; either he will be disgusted with it or he will amuse himself with it as a game and arrive at a state of mind analogous to that of the Greek sophists.
...Henri Poincare

Can one ever understand a theory if one builds it up right from the start in the definitive form that rigorous logic imposes? No; one does not really understand it ... one retains it only by learning it by heart.
...Henri Poincare

I am led to the idea which in substance is that the ordinary first course in calculus should be more largely graphical. The question that the beginner desires answered is what is the

calculus about—what does it do?... Let us be satisfied with a rough definition of the derivative, allowing intuition and not rigor to be the chief factor in supplying it and therefore drawing very largely on the geometrical side.
...W. B. Ford

Elementary mathematics ... must be purged of every element which can only be justified by reference to a more prolonged course of study. There can be nothing more destructive of true education than to spend long hours in the acquirement of ideas and methods that lead nowhere.... [The] elements of mathematics should be treated as the study of fundamental ideas, the importance of which the student can immediately appreciate; ... every proposition and method which cannot pass this test, however important for a more advanced study, should be ruthlessly cut out.
...Alfred North Whitehead

The solution I am urging is to eradicate the fatal disconnection of topics which kills the vitality of our modern curriculum. There is only one subject matter, and that is Life in all its manifestations. Instead of this single unity, we offer children Algebra, from which nothing follows.
...Alfred North Whitehead

Civilization advances by extending the number of operations we can do without thinking about them. Operations of thought are like cavalry charges in a battle—they are strictly limited in number, they require fresh horses and must be made only at decisive moments.
...Alfred North Whitehead

There is a real hypocrisy, quite frequent in the teaching of mathematics. The teacher takes verbal precautions, which are valid in the sense he gives them, but that the students most assuredly will not understand the same way.
...Henri Lebesgue

Unfortunately competitive examinations often encourage [an educational] deception. The teachers must train their students to answer little fragmentary questions quite well, and they give them model answers that are often veritable masterpieces and that leave no room for criticism. To achieve this, the teachers isolate each question from the whole of mathematics and create for this question alone a perfect language without bothering about its relationships to other questions. Mathematics is no longer a monument but a heap.
...Henri Lebesgue

If the limit of the [polygonal areas] had been designated as the "tarababump" of the circle, one would not be permitted to derive from it the value of the tarababump of the sector.... We allow ourselves to do it because instead of the word tarababump, we used the word area!.... Imagine the inevitable confusion that will be caused by making students identify this new area with the areas to which they are accustomed.
...Henri Lebesgue

From the beginning Nature has led the way and established the pattern which mathematics, the language of nature, must follow.
...George D. Birkhoff

Personally, I am always ready to learn, although I do not always like being taught.
...Winston Churchill

The talent [of being able to rapidly make very complicated calculations in the head] is, in reality, distinct from mathematical ability. Very few known mathematicians are said to have possessed it.
...Jacques Hadamard

Why Was Calculus&*Mathematica* Written? What Principles Underlie the Course?

I insist that words are totally absent from my mind when I really think. . . . Even after reading or hearing a question, every word disappears at the very moment I am beginning to think it over; words do not reappear in my consciousness before I have accomplished or given up the research . . . and I fully agree with Schopenhauer when he writes, "Thoughts die the moment they are emodied by words." I think it is also essential to emphasize that I behave in this way not only about words, but even about algebraic signs. I use them when dealing with easy calculations; but whenever the matter looks more difficult, they become too heavy baggage for me.
. . .Jacques Hadamard

The words or language, as they are written or spoken, do not seem to play any role in my mechanism of thought. The psychical entities which seem to serve as elements in thought are certain signs and more or less clear images which can be "voluntarily reproduced and combined."
. . .Albert Einstein

At a great distance from its empirical source . . . , a mathematical subject is in danger of degeneration. At the inception the style is usually classical; when it shows signs of becoming baroque, then the danger sign is up.
. . .John von Neumann

[Mathematics] must remain . . . as a unified vital strand in the broad stream of science and must be prevented from becoming a little side brook that might disappear in the sand.
. . .Richard Courant

[The student] refuses to be bored by diffuseness and general statements which convey nothing. . . . [The student] will not tolerate a pedantry which makes no distinction between the essential and the non-essential, and which for the sake of a systematic set of axioms, deliberately conceals the facts to which the growth of the subject [of calculus] is due.
. . .Richard Courant

The purpose of computing is insight, not numbers.
. . .Richard Hamming

In mathematics, you can be as formal and rigorous as you want to be; but without insight, you'll get nowhere.
. . .Richard Duffin

Many of the math books . . . are full of . . . nonsense—of carefully and precisely defined words that are used by pure mathematicians in their most subtle and difficult analyses, and are used by nobody else. . . . The real problem in speech is not precise language. The problem is clear language.
. . .Richard Feynman

It has been an acknowledged fact, since Poincare pointed an accusing finger at the Twentieth Century, that much of the mathematics of our time has had negative overtones.
. . .Gian-Carlo Rota

Mathematics today is a vital, vibrant discipline composed of many parts which in mysterious ways influence and enrich each other. . . . It is beginning to emerge from a self imposed isolation [from its sister disciplines] and listen with attention to the voices of nature.
. . .Mark Kac

Many conventional academic skills amount to the ability to select and apply ... procedures rapidly and correctly.... [Computer] courseware can concentrate on one skill at a time, in a manner impossible for a textbook and hardly available to the classroom teacher, namely by asking the student to handle only that part of a procedure on which pedagogical stress is to be laid, while other aspects of the same procedure are handled automatically by the computer. This scheme, which combines student interaction with computer assistance, has the merit of focusing attention on the key strategic and conceptual decisions needed to handle a problem.... This should be of significance to both the strong student ... and the weak student.
...Jacob T. Schwartz

When I started off doing mathematics, I wasn't very good at it. I never learned my multiplication tables and it was certainly the conclusion of my teachers at that time that there was no way I would ever go on and do anything ... mathematically oriented. As it turned out, I found out about computers and found out that you could make computers do these kinds of things.
...Stephen Wolfram

Mathematical concepts may be communicated easily in a format which combines visual, verbal, and symbolic representations in tight coordination.
...Ralph Abraham

Before Gutenburg, illustration and type were one in the same; they were inseparable. But afterward, the two disciplines became separate and diverged. Now that we've got the [graphic computer], I can see a medium where they come back together again.
...Scott Kim

I never learned to use a computer myself.... Then when the [Macintosh with Mathematica*] came, I decided this is the machine for me. So I have learned to use it and I depend on the machine heavily.*
...Donald Glaser

Perhaps, ... one should consider teaching mathematics backwards, that is, teaching the images and patterns first and teaching the conventional symbol system, rules, and rigorous processes later. This kind of fundamental change ... may be worth the trouble, especially if it can be shown that some of those who could not do well with elementary mathematics may still do superbly well with the more advanced forms. It may require entirely different ways of teaching, but it may, in the end, be more appropriate for a new era in mathematics and work, when all the easy things will be done by machine and the hard things will be the only things of value left for humans to do.
...Thomas West

There's a terrible problem that I run into in teaching, which is that when you tell people something, you keep them from knowing it. If they find it on their own, they'll know it in a way they never will if you tell them. What I try to do more and more is to bring students to my studio and get them really working.
...Richard Benson

Calculus&*Mathematica* is the authors' best attempt at incorporating all this wisdom into a calculus course.

Acknowledgments

The authors, Bill Davis, Horacio Porta and Jerry Uhl, wholeheartedly thank the persons whose ideas and actions have been essential to Calculus&*Mathematica*.

Calculus&*Mathematica* students at the University of Illinois and Ohio State University for leading the way on the first calculus adventure in three hundred years. Over a five-year period, these students played an essential role in helping us to develop Calculus&*Mathematica*. They taught us that the modern student does have math passions and can really turn on to a calculus course. They taught us what will work and what won't work. They taught us how to write this course. The course benefitted greatly from their ideas about how it should be run. In fact, the Literacy Sheet component of Calculus&*Mathematica* owes its existence to student input. At the times when our calculus batteries needed to be recharged, all we had to do was to look at student work or to sit down and talk to students about their work while they were doing it. The way Calculus&*Mathematica* students throw themselves into their work is our continual inspiration.

Don Brown, coauthor of the preliminary version of Calculus&*Mathematica*, for giving us the confidence we needed in the initial phase of the project. As an Illinois undergraduate physics major, he joined the project at the very beginning to provide the computer know-how needed to transform the project from vaporware to reality. The original versions of the fonts and the *Mathematica*-to-TeX printing programs are all results of his powerfully fertile mind. The project could not have gotten off the ground without the help of this friend.

Alan DeGuzman, **Justin Gallivan**, and **David Wiltz** for forming the original Calculus&*Mathematica* Development Team. All three of these fellows were Calculus&*Mathematica* students in the first full year of the project who stepped forward to share their expertise with us. They taught us how to set up and run a lab, and they have run the technical phase of the project for three years. Their contribution of intellectual energy and creativity was essential to the project. Nearly ready to get their undergraduate degrees, they are all fellows who will go very far in their own endeavors. When they leave, we'll miss them.

Gary Binyamin, **Martha Grover**, **David Kirzner**, **Corey Mutter**, **David Taubenheim**, and **Jenny Welch** for joining the Development Team. Operating under

Acknowledgments 81

the tutelage of the Original Development team, they are all making their own fundamental contributions to Calculus&*Mathematica* and will continue to do so until they graduate.

Thea Colwell, **Ben Halperin**, and **Chris Zeller** for always being there when we needed them.

Chintan Amin, **Mikal Arneborn**, **Lorien Ryan**, **Dave Banyots**, **Steve Casburn**, **Brigitte Blaney**, **Karl Berghauer**, and **Neil Mickelson** for plenty of good advice and for being exemplary lab instructors.

Glenn Scholebo, master typesetter, for his judgment, artistry, and energy in converting the electronic course into print in both the preliminary edition and this edition. Glenn is the main reason the printed supplement to the electronic course is so readable. It's a real pleasure to work with an authentic expert and gentleman.

Beate Zimmer for overall consulting and extraordinary proofreading. When one of us was forced to the sidelines by bad health, she stepped in to keep the project on schedule. She is our friend.

Stephen Wolfram, principal author of *Mathematica*, for calling us over to his office in 1988 and giving us his ideas for what he called "a calculus book in the form of *Mathematica* Notebooks." Although we didn't know it at the time, this was the beginning of the Calculus&*Mathematica* project. We thank him also for his encouragement and support in every phase of the whole project. He understood what we were trying to do before we did. We think Stephen is one heck of a guy.

Francis Sullivan, of the Institute for Defense Analysis, for his advice and direction both mathematical and nonmathematical. He encouraged us to start the project and has been a continual source of wisdom for the project. Among many other things, he helped us to understand the difference between traditional classroom calculus and honest calculus as practiced outside the traditional classroom. We think Francis is also one heck of a guy.

Bruce Carpenter for understanding Calculus&*Mathematica* better than the authors understand it. He is a man of incredible originality, insight, and drive; we are happy to have him on our team.

Theodore Gray for inventing the Notebook Front End of *Mathematica*, to which this course is a response.

Dan Grayson, coauthor of *Mathematica*, for introducing us to *Mathematica* and for plenty of pithy comments.

Allan Wylde, our original Addison-Wesley publisher, editor, and visionary who went so far as to back our project with badly needed financial and personal support at a time when most everyone was skeptical about the project. The day he left Addison-Wesley was a sorry day for us.

Jerome Grant, our Addison-Wesley editor, who surprised us by proving to be a math editor with brains, vision, and drive. He gave us the freedom we never expected to get. We like this guy.

Richard Wilson, associate chancellor of the University of Illinois at Urbana-Champaign, for his unflagging support from the very beginning. Without this friend, we would have been dead in the water.

Ward Henson of the University of Illinois for saving this project from an early death during his term as chair of the mathematics department at Illinois.

Joan Leitzel who, during her term as associate provost at Ohio State University, gave the support necessary for establishing the early Calculus&*Mathematica* labs there.

Acknowledgments

The University of Illinois at Urbana-Champaign and **Ohio State University** for giving us the freedom to develop, revise, and teach Calculus&Mathematica in their classrooms.

Amy Young, member of the first-ever Calculus&Mathematica class and now with Wolfram Research, for her uniquely incisive advice and support from the very beginning. She is a very special friend.

All the other folks at Wolfram Research Inc. for unwavering support of the project. We thank especially **Shawn Sheridan**, **Glenn Scholebo**, **Prem Chawla**, **Scott May**, **Tom Wickham-Jones**, **Paul Katula**, **Kevin MacIsaac**, **Eric Blankenburg**, **Paul Abbott**, **Dara Pond**, **Zoe Midler**, **Lisa Shipley-Jones**, **Jane Rich**, **Laurie Gilmore**, **Jerry Keiper**, **Dan Lichtblau**, **Ben Friedman**, **Brad Horn**, **Tom Sherlock**, **Bo Liu**, **Joe Grohens**, **Jamie Peterson**, **John Cochran**, **Melissa Idleman**, **Christy Uden**, **Joe Kaiping**, **David Withoff**, and **John Bonadies**, who all did things for us that they didn't really have to do.

Andy Fisher, **Karen Wernholm**, **Peter Blaiwas**, and **Sean Angus** of Addison-Wesley, for going about their business with cheer and aplomb.

Tony Peressini, at Illinois, for lots of advice and support. Every project like this needs a steady source of practical wisdom; he is ours. His conversion from the traditional course to Calculus&Mathematica was a signal event for us. Tony's support of the project through the Illinois High School Distance Calculus Network, which he heads, has been vital. He is a very special friend.

Alayne Parson and **Tom Ralley** at Ohio State for their enthusiastic support for the project. They have, at times, been more optimistic than we have about the future of the course and, in particular, in the pedagogical tenets upon which it is based.

Paul McCreary for sharing his ideas on group learning and for proving that Calculus&Mathematica can be a beneficial experience for some students who would normally be considered to be at risk in a calculus course.

Dana Scott for his belief in the project and his efforts on our behalf.

Jonathan Manton, **Eric Hjelmfeldt**, and **Soren Lundsgaard** for help in the very early stages of Calculus&Mathematica at Illinois.

The National Science Foundation for grants for development.

Apple Computer for machine support and for making two videos about the project. **John Noon** for convincing Apple Computer that this project should be supported.

Ron Weissman and **David Spitzler** of NeXT Computer for loaner computers and for their positive attitude about the project.

Bernard Madison for his early recognition of the merits of the project and for writing the first accurate, succinct description of Calculus&Mathematica.

Emily Mann Peck, associate dean of liberal arts and sciences at Illinois, for her support.

George Badger, director of the University of Illinois Computer Services Office, for cooperation far beyond what we expected or deserved.

Charlie Bender, director of the Ohio Supercomputer Center, and director of academic computing at Ohio State, for getting the project off the ground there by suggesting the construction of a computer lab usable for Calculus&Mathematica, and for letting **Holly Hirst**, then his assistant, teach in the first year of the project.

Acknowledgments

Bob Dixon, **Steve Gordon**, **Randy Jackson**, **Terry Reeves**, and **Jeff Schluep** of Academic Computing Services at Ohio State for support, installation, and systems support for the various labs that have been homes to Calculus&*Mathematica*.

Juan Manfredi, of the University of Pittsburgh, for his continuing support. When he told us he liked the project enough to teach it, we began to feel that it could become real.

Elias Saab, chair of the mathematics department at the University of Missouri, Columbia, for proving that a large-scale implementation of Calculus&*Mathematica* can be successful.

Russell Howell, **Dennis Schneider**, **Juan Manfredi**, **Lew Lefton**, **Enid Steinbart**, **Ken Holladay**, **Elias Saab**, **Dana Weston**, **Carruth McGehee**, **Neil Stoltzfus**, **Paul Wellin**, **Bill Emerson**, **Louis Talman**, **Dan Yates**, **Tony Peressini**, **Don Sherbert**, **Peter Loeb**, **Elliott Weinberg**, **Bob Muncaster**, **Alice Iverson**, **Barbara Beechler**, **Mel Henrickson**, **Rob Beezer**, **Barry Turett**, **Jack Nachman**, **Elizabeth Covington**, **Robin Sanders**, **Fred Andrew**, and **Tom Morley** for giving Calculus&*Mathematica* an early try in their university and community college calculus courses. They proved that Calculus&*Mathematica* can ignite students' mathematical passions in a variety of settings.

Judy Holdener, **Janice Malouf**, **Bruce Carpenter**, **Bill Hammack**, **Eric Hjelmfeldt**, **David Ose**, **Judy Walker**, **Todd Will**, **Heather Hulett**, **Tammy Hummel**, **Myung Sook Chung**, **Gregory Michalopoulos**, and all the other Illinois teaching assistants and professors who were game enough to run early sections of Calculus&*Mathematica* as it was being developed.

Holly Hirst, **Tom Ralley**, **Alayne Parson**, **Amy Lee**, **Cheryl Stitt**, **Ed Overman**, **Steve Giust**, **Jere Green**, **Marc McClure**, **Jim Brazelton**, and all the other Ohio State teaching assistants and professors who were also game enough to run early sections of Calculus&*Mathematica* as it was being developed.

Jeff Hirst for sharing his perspective on the course and for his support for Holly while we struggled through the first year of learning about the course.

David Appleyard for letting his ideas rub off on us during the semester he spent in the Calculus&*Mathematica* lab at Illinois.

Francis Sullivan, **Karen Uhlenbeck**, **Luis Caffarelli**, **Albert Fassler**, "Spud" **Bradley**, **Martin Flashman**, **Deborah Hughes Hallett**, **Bernard Madison**, **Cleve Moler**, **Peter Lax**, **Peter Ponzo**, **Dana Scott**, **Gil Strang**, **David Tall**, **Edward Tufte**, **Mac Van Valkenberg**, **Ken Wilson**, **Jerry Johnson**, and **Stephen Wolfram** for taking time out of their own schedules to visit the lab as Calculus&*Mathematica* was being developed.

Shirley Treadway of Robinson High School for suggesting that Calculus&*Mathematica* could be used by high school students in distance education.

Shirley Treadway, **Kay Hall**, **Deana Brashear**, **Jackie Wood**, **Mark Catt**, and **Mark Calvert** for sponsoring Calculus&*Mathematica* at their high schools and for helping to prove that Calculus&*Mathematica* can be used successfully in distance education.

Nora Sabelli, **Lisa Bienvue**, and **John Duban** of the National Center for Supercomputer Applications (NCSA) for help in networking for distance education.

Deborah Crocker for the first systematic study of the learning styles and characteristics of Calculus&*Mathematica* students in her doctoral dissertation at Ohio State University. We were very pleased with her conclusion that one of the striking differences between Calculus&*Mathematica* students and traditional students is that Calculus&*Mathematica* students believe they can attack and solve problems that require the use of calculus. We were glad to have her further evidence that Calculus&Mathematica students develop a stronger conceptual understanding of calculus than their peers in traditional courses.

Kyung Mee Park for the first systematic comparison of Calculus&Mathematica students against traditionally trained students in her doctoral dissertation at the University of Illinois. We were very gratified by her conclusions that Calculus&Mathematica students are considerably more likely to try multiple approaches to a problem than students from the traditional course, that Calculus&Mathematica students cannot be distinguished from students in the traditional course on the basis of their ability at hand calculation, and that Calculus&Mathematica students demonstrate considerably richer ability to identify relations among calculus ideas than students in the traditional course.

Ken Travers for being the director of Kyung Mee Park's dissertation research and for being a steady source of calm wisdom. His continued interest in the course has been much appreciated.

Albert Fassler for his work in transforming Calculus&Mathematica for the Swiss classroom and for ringing the Calculus&Mathematica bell in Europe.

William Graves, **Ladnor Geissinger**, **Lester Senechal**, and **Gerald Porter** for selecting Calculus&Mathematica as the prototype interactive text for the national IBM-MAA-NSF Interactive Mathematics Text Project.

Peter Lax for issuing the first call for calculus revolution by proclaiming that "calculus as currently taught is full of inert topics."

Ralph Abraham for his interest in the project and for his view that "mathematical concepts may be communicated easily in a format which combines visual, verbal, and symbolic representations in tight coordination." He has been at this business a lot longer than we have. We hope he likes what he sees.

Lynn Arthur Steen for smelling a course like Calculus&Mathematica in the wind even before we began to work on it.

The authors of the National Research Council reports *Everybody Counts* and *Moving Beyond Myths* for issuing a mandate for a course like Calculus&Mathematica. In a very real way, Calculus&Mathematica is a reaction to these reports. In fact, we believe Calculus&Mathematica to be in line with every recommendation in these reports—from the electronic text to "mathematics as it really is" and to classes without lectures.

David Tall for papers on using computers for visualization in the mathematics classroom. His ideas opened our eyes to what is possible. His visit to our lab gave us a jump start.

Thomas West for sharing his ideas on visual learning in a long conversation and for writing a manifesto for visual learning in his book, *In the Mind's Eye: Visual Thinkers, Gifted People with Learning Disabilities, Computer Images, and the Ironies of Creativity*. His ideas had a profound effect on the revision of Calculus&Mathematica from the preliminary to the present edition.

Andrew Gleason for reminding us over and over that common sense is the great engine for driving mathematics.

Henry Edwards for his interest in this project and for writing *The Historical Development of Calculus*, which gave us many provocative morsels to savor when we were thinking about what calculus is.

Deborah Hughes Hallett, honcho of the Harvard Calculus Consortium, for her encouragement and for hour after hour of sagacious advice on what should be in, what should be out, what will work, and what will not work. She is one of the rare individuals who understands that mathematics itself must be rethought before better calculus courses can begin. She has been a behind-the-scene advisor to the project almost from the beginning. Our special thanks go to her.

Jim Callahan, **Ken Hoffman**, **Donal O'Shea**, and **Lester Senechal**, of the Five Colleges Calculus Project, for showing us what calculus reform really means. Many of their ideas are reflected in Calculus&*Mathematica*. They have rethought the mathematics of calculus; their view of calculus remains a beacon for those who question the traditional course.

Martin Flashman for sharing his ideas for a sensible calculus with us in a visit to our lab. His ideas have influenced this course even more than even he might believe.

Ron Douglas, **Al Tucker**, and **Tom Tucker** for some profound comments about the role of algebra, trigonometry, and technology as delivery vehicles in modern calculus.

James Glimm for expressing his view that the derivative should be introduced as a measurement of the growth rate rather than the slope of a tangent line. His views confirmed our belief in calculus as a course in measurements. We also thank him for his view that computers, not calculators, are the correct technology for the modern calculus classroom.

Stephen Jay Gould for reminding us, through his writings, that a course like this cannot ignite students' mathematical passions if any part of it is copied or adapted from an existing text.

Edward Tufte for his books *The Visual Display of Qualitative Information* and *Envisioning Information*.

Igor Kluvanek, our old and dear late friend, for his papers "What's Wrong with Calculus?" and "Archimedes Was Right." It's hard to accept that we will never again have the opportunity to sit down with him with a good drink and talk about the central issues of mathematics and of life.

Gil Strang for his book *Introduction to Applied Mathematics*. This book had a decisive effect on our ideas about vector calculus as a course about measurements rather than formulas.

J. D. Murray for his book *Mathematical Biology*. This book helped us to make the decision to use biological models to illustrate the meaning of the derivative in early calculus.

Jacob T. Schwartz for his view that "courseware can concentrate on one skill at a time, in a manner impossible for a textbook and hardly available to the classroom teacher, namely by asking the student to handle only that part of a procedure on which pedagogical stress is to be laid, while other aspects of the same procedure are handled automatically by the computer."

Emil Artin for his Princeton calculus lectures. His course proved that a strong calculus course need be neither formal nor laden down with heavy terminolgy or algebra. We also thank him for his position that when a mathematical idea has a geometric interpretation, then that idea should be introduced via geometry instead of algebra.

Blaise Pascal, **Gottfried Wilhelm Leibniz**, and **Augusta Ada Byron** for calling early attention, in their writings, to the fact that machine calculations have an essential role in mathematics.

Leonhard Euler for his book *Introduction to the Analysis of the Infinite*, which showed us a style of writing appropriate for *Mathematica* Notebooks.

Henri Poincare for arguing, in his writings, in favor of mathematics courses that emphasize insight over rigor. In doing the mathematical archeology required for a new calculus course, we came to the view that our visual approach was first advanced by

Poincare. In fact, we go so far as to say that the traditional calculus and differential equations courses are courses in pre-Poincare mathematics. We offer Calculus&Mathematica as our best attempt at a post-Poincare calculus course.

Alfred North Whitehead for announcing, in his writings, a general principle for determining what parts of a mathematics course are inert and should be dropped. We also thank him for his view that "civilization advances by extending the number of operations we can do without thinking about them."

Henri Lebesgue for his writings on the pedagogy of the integral. He shared Artin's idea that instead of defining area by an integral, the integral should be introduced as a device to measure area. It was his idea to deemphasize the indefinite integral (the integral without limits) in calculus. It was also his idea to emphasize the decimal form of a number.

Richard Feynman for his view that most mathematics textbook authors are wrong when they confuse clarity with precision.

Many other persons, living or dead, for contributing to Calculus&Mathematica by sharing wisdom and problems through direct suggestion or in their writings. Their specific contributions are mentioned as appropriate in the main body of the course. We thank them.